Control of Quantum-Mechanical Processes and Systems

Mathematics and Its Applications (*Soviet Series*)

Volume 56

Control of Quantum-Mechanical Processes and Systems

by

A. G. Butkovskiy
Laboratory for Distributed Parameter Systems,
Institute of Control Sciences, Moscow, U.S.S.R.

and

Yu. I. Samoilenko
Ukrainian Academy of Sciences, Control Systems Department,
Institute of Cybernetics, Kiev, U.S.S.R.

KLUWER ACADEMIC PUBLISHERS
DORDRECHT / BOSTON / LONDON

Library of Congress Cataloging in Publication Data

Butkovskiĭ, A. G. (Anatoliĭ Grigor'evich)
    [Upravlenie kvantovo-mekhanicheskimi protsessami. English]
    Control of quantum-mechanical processes and systems / by A.G.
Butkovskiy and Yu.I. Samoilenko.
        p.   cm. -- (Mathematics and its applications ; v. 56)
    Translation of: Upravlenie kvantovo-mekhanicheskimi protsessami.
    Includes bibliographical references and index.
    1. Control theory.   2. Quantum theory.   I. Samoĭlenko, I͡U.
I. (I͡Uriĭ Ivanovich)   II. Title.   III. Series: Mathematics and its
applications (Kluwer Academic Publishers) ; v. 56.
QA4023.B8313 1990
629.8'312--dc20   90-4706

ISBN-13: 978-94-010-7392-9   e-ISBN-13: 978-94-009-1994-5
DOI: 10.1007/978-94-009-1994-5

Published by Kluwer Academic Publishers,
P.O. Box 17, 3300 AA Dordrecht, The Netherlands.

Kluwer Academic Publishers incorporates
the publishing programmes of
D. Reidel, Martinus Nijhoff, Dr W. Junk and MTP Press.

Sold and distributed in the U.S.A. and Canada
by Kluwer Academic Publishers,
101 Philip Drive, Norwell, MA 02061, U.S.A.

In all other countries, sold and distributed
by Kluwer Academic Publishers Group,
P.O. Box 322, 3300 AH Dordrecht, The Netherlands.

Printed on acid-free paper

Translated from the Russian by M. B. Burov

This is the translation of the original work
УПРАВЛЕНИЕ КВАНТОВОМЕХАНИЧЕСКИМИ ПРОЦЕССАМИ
Published by Nauka Publishers, Moscow, © 1984.

SERIES EDITOR'S PREFACE

Mathematics is a tool for thought. A highly necessary tool in a world where both feedback and non-linearities abound. Similarly, all kinds of parts of mathematics serve as tools for other parts and for other sciences.

Applying a simple rewriting rule to the quote on the right above one finds such statements as: 'One service topology has rendered mathematical physics ...'; 'One service logic has rendered computer science ...'; 'One service category theory has rendered mathematics ...'. All arguably true. And all statements obtainable this way form part of the raison d'être of this series.

This series, *Mathematics and Its Applications*, started in 1977. Now that over one hundred volumes have appeared it seems opportune to reexamine its scope. At the time I wrote

> "Growing specialization and diversification have brought a host of monographs and textbooks on increasingly specialized topics. However, the 'tree' of knowledge of mathematics and related fields does not grow only by putting forth new branches. It also happens, quite often in fact, that branches which were thought to be completely disparate are suddenly seen to be related. Further, the kind and level of sophistication of mathematics applied in various sciences has changed drastically in recent years: measure theory is used (non-trivially) in regional and theoretical economics; algebraic geometry interacts with physics; the Minkowsky lemma, coding theory and the structure of water meet one another in packing and covering theory; quantum fields, crystal defects and mathematical programming profit from homotopy theory; Lie algebras are relevant to filtering; and prediction and electrical engineering can use Stein spaces. And in addition to this there are such new emerging subdisciplines as 'experimental mathematics', 'CFD', 'completely integrable systems', 'chaos, synergetics and large-scale order', which are almost impossible to fit into the existing classification schemes. They draw upon widely different sections of mathematics."

By and large, all this still applies today. It is still true that at first sight mathematics seems rather fragmented and that to find, see, and exploit the deeper underlying interrelations more effort is needed and so are books that can help mathematicians and scientists do so. Accordingly MIA will continue to try to make such books available.

If anything, the description I gave in 1977 is now an understatement. To the examples of interaction areas one should add string theory where Riemann surfaces, algebraic geometry, modular functions, knots, quantum field theory, Kac-Moody algebras, monstrous moonshine (and more) all come together. And to the examples of things which can be usefully applied let me add the topic 'finite geometry'; a combination of words which sounds like it might not even exist, let alone be applicable. And yet it is being applied: to statistics via designs, to radar/sonar detection arrays (via finite projective planes), and to bus connections of VLSI chips (via difference sets). There seems to be no part of (so-called pure) mathematics that is not in immediate danger of being applied. And, accordingly, the applied mathematician needs to be aware of much more. Besides analysis and numerics, the traditional workhorses, he may need all kinds of combinatorics, algebra, probability, and so on.

In addition, the applied scientist needs to cope increasingly with the nonlinear world and the extra mathematical sophistication that this requires. For that is where the rewards are. Linear models are honest and a bit sad and depressing: proportional efforts and results. It is in the non-linear world that infinitesimal inputs may result in macroscopic outputs (or vice versa). To appreciate what I am hinting at: if electronics were linear we would have no fun with transistors and computers; we would have no TV; in fact you would not be reading these lines.

There is also no safety in ignoring such outlandish things as nonstandard analysis, superspace and anticommuting integration, p-adic and ultrametric space. All three have applications in both electrical engineering and physics. Once, complex numbers were equally outlandish, but they frequently proved the shortest path between 'real' results. Similarly, the first two topics named have already provided a number of 'wormhole' paths. There is no telling where all this is leading - fortunately.

Thus the original scope of the series, which for various (sound) reasons now comprises five subseries: white (Japan), yellow (China), red (USSR), blue (Eastern Europe), and green (everything else), still applies. It has been enlarged a bit to include books treating of the tools from one subdiscipline which are used in others. Thus the series still aims at books dealing with:

- a central concept which plays an important role in several different mathematical and/or scientific specialization areas;
- new applications of the results and ideas from one area of scientific endeavour into another;
- influences which the results, problems and concepts of one field of enquiry have, and have had, on the development of another.

Control theory is concerned with equations of the form $\dot{x} = f(x,u)$ where x models the state of some system, for instance a quantum mechanical system, and the u are control (input) parameters which can be used to try to make the system behave in some desired manner. 'Classical' control theory is a vast, and vastly important, subject (as can be readily imagined). Here 'classical' is used as opposed to 'quantum', for although the beginnings of (automatic) control devices go back to antiquity (cf. e.g. 'The Book of Ingenious Devices') the subject has only comparatively recently really flowered; principally in the last 30 years.

In terms of control theory the control of quantum systems involves the theory of infinite dimensional control systems with distributed parameters, but there are, of course, special properties which can be exploited.

Recent advances in technology, particularly in lasers, have made the detailed control of quantum mechanical systems possible (at the quantum level), and this is a first systematic book on the topic.

Besides a comprehensive treatment of quantum (optimal) control it offers a set of real applications to systems with negative susceptibility and conductivity.

The shortest path between two truths in the real domain passes through the complex domain.

> J. Hadamard

La physique ne nous donne pas seulement l'occasion de résoudre des problèmes ... elle nous fait pressentir la solution.

> H. Poincaré

Never lend books, for no one ever returns them; the only books I have in my library are books that other folk have lent me.

> Anatole France

The function of an expert is not to be more right than other people, but to be wrong for more sophisticated reasons.

> David Butler

Bussum, 25 April 1990   Michiel Hazewinkel

Contents

Preface

Control of quantum-mechanical processes and systems is an ever more topical problem. Purposeful control of the quantum state of matter has long since been the focus of investigation for many researchers working in various branches of physics. The advent of lasers and laser technology gave a new push to the development of control problems because their solution required designing automatic systems with feedback and special software.

These problems are primarily associated with the development of control systems needed to optimize and stabilize quantum generators in order to achieve the minimum of their operation cost and the maximum accuracy while at the same time expanding the range of adjusted parameters. On the other hand, coherent radiation impact on matter opens up new vistas in technology. It offers a means of carrying out purposeful transformations of matter while changing its composition, structure or state (the 'coherent technology'). In terms of control, this can be regarded as the problem of producing a field control (in particular, electromagnetic) or a corpuscular action influencing systems of atoms or molecules in order to obtain required states or structures on the microlevel.

This book sums up the results of modern research in the theory of automatic control, which can be used to solve many theoretical problems of control systems defined by the equations of quantum physics. In terms of control, any such problem is related to the theory of control systems with distributed parameters, where one deals with the processes of control by means of various physical fields and with the principles of designing both open-loop and closed-loop systems (with feedback) to control the distributed parameter systems.

Chapter 1 offers mathematical formulations of control problems with due regard for the special features of quantum systems. These are problems of optimal control, parametric optimization, finite control, control of the spectrum of operator eigenvalues, control of the averages and variances of physical variables or, in a more general setting, control of the moments and other functionals of given physical quantities varying under the effect of external control factors, generally fields of various physical natures. Usually these are macroscopic electromagnetic fields, for example, electric or magnetic ones. The problems are considered in terms of pure states defined by wave functions, and mixed states defined by statistical operators.

Particular emphasis has been placed on the problem of *controllability of a quantum process*, which is important both in principle and from the practical viewpoint. This problem is posed in Chapters 2 and 3 in various terms. One of the formulations is given in terms of the vector function $\psi(t)$ of a quantum state and consists in the following. Given an initial state $\psi(0) = \psi_0$, find the admissible controls (for example, variation in the strength of an electric or magnetic field or both) such that at an arbitrary or specified instant T the system will be in the desired final state $\psi(T) = \psi_1$. A similar problem can be formulated in terms of non-stationary operators of physical variables, statistical operators, distribution moments of statistical ensembles, etc.

This problem has two aspects. One of them is the identification of a number of necessary and sufficient conditions for the solution of the problem. Verification of these conditions guarantees the attainability or unattainability of the desired states in principle. The other aspect of the controllability problem is the actual synthesis of exact or approximate analytic or numerical controls realizing the control of the system (the achievement of the desired state). In terms of the general theory of control, this is the *problem of finite control*, which is often very troublesome. Any practical implementation of finite control of particle ensembles, taking into account both statistical and quantum effects, is closely associated with the problem of producing control fields with required characteristics.

Very often there are many possible solutions of the finite control problem. It is then worth while posing an additional problem: select from the set of all possible finite controls, one that minimizes a given loss functional (or maximizes a desired quality criterion), that is, formulate a problem of optimal control. As to the loss functional, it can be one describing the productivity of a technology, its energy or mass consumption, or a more general 'economical' functional of the cost of control.

Chapter 4 deals with various aspects of optimal control of quantum-mechanical systems. Much thought is given to algebraic constructions (in particular, those derived from group theory) used in solving optimization problems.

A special place in the book is set aside for the presentation of the theory of negative quasistatic susceptibility and its diverse applications (Chapters 5 and 6). The problems of this theory tie up with the rest of the book at least in that the attainment of a negative static susceptibility state should be regarded in a consistently developed theory as both the purpose and the means of control of a quantum system state.

Appendices 1–3 are concerned with the construction of mathematical models of quantum systems, the theory of finite control and the analogy between continuous media and controlled finite-dimensional dynamical systems.

Acknowledgements
The authors have been greatly helped in the preparation of the Russian manuscript by N.L. Lepe, E.I. Pustylnikova, O.A. Khorozov, and A.P. Andrushchenko. We extend our deep gratitude to all of them.

A. Butkovskiy, Yu. Samoilenko

CHAPTER 1

The Problem of Control on the Quantum Level

1.1. Introduction

Purposeful control of quantum physical processes has long since been the subject of many investigations. Problems of controlling microprocesses and quantum ensembles were posed and solved for plasma and laser devices, particle accelerators, nuclear power plants, and units of automation and computer technology (see Nikityuk, 1967; Venikov, 1976; Petrov et al., 1976). Although no quantum transition based processors have so far (by 1987) been developed, there is no doubt that their production is near at hand. Some quantum effects have already been applied in optoelectronic computer devices, but their integration into diverse computer systems still belongs to the future.

There has recently been an intense interplay between methods of quantum physics and the theory of control, although the problem of controlling quantum states was in fact brought forth at the rise of quantum mechanics. For instance, many experimental facts of quantum mechanics were established with the use of macroscopic fields acting on quantum ensembles, which from the modern viewpoint can be regarded as control. As the technology of experiment evolved, new problems in controlling quantum systems arose (Waugh, 1976; Samoilenko, 1972, Samoilenko and Khorozov, 1980; Butkovskiy and Samoilenko, 1979a), and their solution required special methods (Andreev, 1982; Brockett, 1973; Butkovskiy and Samoilenko, 1979a; Samoilenko, 1982a, 1982b). The application of non-perturbative methods of measurement (Kholevo, 1980; Moncrief, 1978; Dodonov et al., 1980) opens up interesting possibilities; in particular those that do not perturb states with definite energy (Vorontsov, 1981).

The objective of control in thermal excitation of quantum systems can be formulated in terms of statistics for large numbers of quantum states, which brings this branch of the quantum theory of control closer to the well-developed theory of statistical process control. Here quantum features reveal themselves in discrete energy spectra and the symmetry or antisymmetry of wave functions for multiparticle systems.

The problem of selective non-thermal control of quantum processes is more intriguing. It can be solved for the combination of highly coherent radiation and a well-defined pattern of the energy spectrum of the material. The invention of lasers was a prerequisite for the advent of techniques of simulation and control needed for nuclear and chemical reactions, transport in various media, biochemical processes on the subcellular level, etc.

It is an exciting prospect to be able to produce novel quantum transition-based elements for control systems and quantum holography and microscopy (for example, in connection with the increasing need for accuracy of measurement in experiments with test bodies (Braginskii, 1970). A number of quantum effects, such as the Josephson effect and the phenomena of induced emission, are employed in controlled units and devices in computer technology. However, so far the problems of controlling quantum processes have been considered mainly in the space of averaged physical variables rather than the space of quantum-mechanical states. This is only valid if the characteristic control time is much greater than the relaxation time of the underlying processes, so that the state variables and controlled processes are identical with the common macroscopical ones, and the quantum origin of the modelling equations is only exhibited in the non-linear dependencies involved. We shall not discuss these cases further but rather start considering the possibility of controlling quantum states in both pure and mixed ensembles.

Bearing in mind that any purposeful influence exerted in order to change the state of a quantum-mechanical object can generally be regarded as control, we see that this problem should encompass the identification and active diagnosis of quantum states. Of great significance now are methods of coherent optical communication, where a consideration of the quantum features of the photon channels is unavoidable. The quantum theory of verification of hypotheses is dealt with in Helström (1976), which includes a detailed bibliography on the subject. The Heisenberg uncertainty relation was suggested for use in estimating the limits of the possibility of controlling microscopic processes and quantum ensembles in Krasovskii (1974), Petrov et al. (1976), and Petrov and Uskov (1975) from the viewpoint of information theory. It should be noted that such an estimate may prove to be too sharp for special classes of states and operators of observables. The theory of optimal quantum estimation is presented in Helström (1976) and in Kholevo (1980).

Interesting control problems are formulated and developed for the optimization of transition processes in the excitation of various media by means of radiation in the optical and other ranges, as well as for the optimization of processes of transforming the modes of oscillation or processes of maintaining desired deformation of wavefronts.

The wide spectrum of the above-mentioned problems can be reduced to the following theoretical schemes:

(1) construction of a set of mixed states accessible from a given initial state;

(2) identification of the set of controls steering the system from a given initial state to a desired accessible state with the greatest or specified probability;

(3) identification of a control that is optimal with respect to a given criterion, e.g. the response time or the minimum of switches (in bang-bang control);

(4) identification of the measuring instrument operator and the method of its implementation by means of programmed control providing the most reliable measurement at a given instant T;

(5) construction of a system of microscopic feedback providing for the possibility of control with accumulation of data.

We shall indicate in this book some principles of reducing these schemes to the ones already known in the theory of control.

Quantum physics and control are two rich and well-developed theories. Their efficient unification requires joint efforts of specialists in both fields. When two such abundant theories are joined, the effect is multiplicative rather than additive because they amplify each other's potential in proportion to their range of development.

Some concepts of quantum and statistical mechanics are fundamental for this book, and the information on them is presented in Appendix 1. Here the authors have mainly followed the well-known textbooks by Davydov (1973) and by Landau and Lifshits (1963, 1976) and monographs by Dirac (1958), Mackey (1962) , and by Neumann (1932). As to the fundamentals of the mathematical theory of control and observation, there also is vast literature on them (Krasovskii and Pospelov, 1962; N. Krasovskii, 1968; Gabasov and Kirillova, 1971; Roitenberg, 1971; Zubov, 1975; A. Krasovskii, 1968; Pugachev, 1957; Pugachev et al., 1974; Feldbaum, 1971; Feldbaum and Butkovskiy, 1971; Pontryagin et al., 1963; Voronov, 1966, 1979; Sirazetdinov, 1977; Lurie, 1975). The specifics of quantum problems of control are discussed by comparing them with similar classical problems. Sometimes this is convenient to carry out in a quasiclassical approximation.

1.2. A Quantum Process as the Object of Control

When a dynamical object of control is considered, the control function or operator is commonly indicated and the objective of control is set as a mathematical requirement on the object's state variables. Besides, measured parameters and their relation to the state variables should be indicated in problems of control with feedback, that is, in the problems of synthesis and identification.

Before we deal with the form of the Hamiltonian defined by the physical nature of a system, let us remark that control can generally be viewed as an explicit dependence of the Hamiltonian on time, or the possibility of varying its 'constant' parameters. The very possibility, if any, of controlling a quantum system, that is, purposely changing its state or structure through macroscopic actions, consists in the system's being able to exchange energy with the environment on either the micro- or macro-interaction level. This can be regarded as the physical nature of control. Inasmuch as the energy of the controlled system varies, an essential inference is that its operator becomes time-dependent.

Macroscopic control of a quantum system can be both direct and indirect. In the former case the system's Hamiltonian is directly varied in time according to the required control. In the latter case the system is acted on, through the interaction Hamiltonian, by a subsystem subjected to macroscopic control and playing the role of a microscopic regulator. Theoretically both variants are very much the same, the difference being purely physical. In order to set up a number of problems, it is sufficient to assume that the average physical variables (the observables) in the mathematical formulation of the objective of control are only presented with the use of operators defined in the subsystem state subspace, and the corresponding part of the Hamiltonian does not depend on time explicitly.

In order to describe the process of measurement, the system's Hamiltonian should be supplemented by the Hamiltonian of the macroscopic instrument (measuring device) whose current state may be considered to be reliably known to the observer. In the microcosm, the process of measurement influences the measured value noticeably. This influence is less if the interaction between the instrument and the measured system is weak. However, in order to decrease the uncertainty in the estimate of the state of the instrument, it is necessary to increase appropriately the period of observation. In principle, if there is a system with complete information, it is possible to establish, by means of long enough observation, which of its stationary states the system is in before the initiation of control.

Therefore, the role of measurement within a system in its pure state can be reduced to the identification of the initial conditions, for which an infinitely large lapse of time in the 'past' is required. Owing to coherent influence of a great number of identical systems on the instrument, the time necessary for the measurement can be made arbitrarily small. In the limit, it proves to be possible to carry out current measurement of the macroscopic averages of various quantities and then use them to restore, partially or in full, the vector state at any instant. This allows us to establish feedback through macroscopic channels.

However, if the interaction between the instrument and the system is strong, the system's pure state is generally destroyed in the process of measurement and becomes mixed. But if the initial state is a mixed one, measurement in the process of control can decrease the original uncertainty. Note that if the measured physical quantity (observable) is defined by an operator commuting with the system's Hamiltonian, the originally determined state cannot be perturbed by any subsequent measurements.

For the measurement of mixed states it is necessary to use an extended model including both the system and the macroscopic instrument. The description of the system can be multilevel (hierarchical), that is, consisting of quantum, quasiclassical, and classical parts. In this connection we have to clarify the problem of joining the subsystems at various levels of description.

The various asymptotic methods, including those employed in boundary-layer theory, can be useful here.

Possible objectives of control of quantum systems can be classified similarly to the general classification in the theory of control. Of primary interest are the problems steering a system from one state to another or taking it through a sequence of required states. This objective is formulated as boundary conditions for the state vector or dynamic operators, which leads to a two-point or a multipoint boundary problem, respectively.

The problems of estimating the parameters of a system or its state in the present from the data of past measurements (identification) or future measurements (observation) (Belavkin, 1975) with the use of controls can also be viewed as problems of control. Physicists have long since been interested in identifying the parameters and the form of the Hamilton function from the data of spectroscopic observations. This is the *inverse spectral problem*: here one is required to recover the operator from a given spectrum.

Another possible independent objective of control is to shape a desired energy spectrum by means of macroscopic control of the system. The objectives of control can also be to stabilize or destabilize some specified states of the system, to attain a strongly non-equilibrium distribution (for example, in a quantum amplifier or generator), to achieve negative quasistatic susceptibility of the medium (see Chapter 4), or to decouple subsystems that interact with each other in their normal state.

An interesting example of non-trivial control of a quantum-mechanical system is the influence of a magnetic field on a system of spins at nuclear magnetic resonance (NMR), the aim being an essential increase in the resolution of NMR (Waugh, 1976). Control in this case is a strictly determined periodic action: rapid rotation of a studied sample in the magnetic field, irradiation by a radio frequency field under a certain ('magical') angle or irradiation by a sequence of powerful impulses. As is shown in Waugh (1976) and Slichter (1980), specially adjusted control of this kind and a special (stroboscopic) technique of observation makes it possible to obtain a Hamiltonian of the system such that internal subsystems become decoupled, which is necessary to reduce the contribution of internal heterogeneities that promote the relaxation of transverse magnetization and lead to undesirable non-uniform broadening of the NMR lines.

An important and interesting problem both in theory and practice is that of attaining prescribed or extremal generally non-equilibrium (statistical) distributions (states) of pure or mixed ensembles of systems. It may be important in practice that certain macroscopic quantities should take specified or extremal values in such states. These states can often be very useful in many areas of physics. For instance, the central problem in quantum magnetometry (Waugh, 1976) is to obtain (at least for a short time) the maximum polarization of microscopic momenta of nuclei, electrons or atoms under study. As a rule, these extremal states and the corresponding non-equilibrium distributions (in the absence of external influences) are currently achieved with the use of extremal external conditions, such as a low or ultralow temperature and the maximum possible external magnetic field strengths.

In consequence, in most cases only stationary or equilibrium distributions are employed and they are controlled by stationary or static fields. However, far from being fully investigated and employed are the possibilities of obtaining the desired or extremal values and corresponding distributions through dynamic control by external fields that vary non-trivially in both space and time. Unlike the commonly used total external control manipulating only the most general properties of the stationary state, the new, finer external control should be based on the finest properties of a quantum system or process and be *coherent* with them. We shall call it *coherent control.*

The objectives of control can be formulated not only as boundary conditions but also as problems of optimal control of the states or functionals. For instance, if a subsystem is a part of an adiabatically isolated system, it is sometimes advantageous to pose the problem of transferring the subsystem to the state with minimum entropy or demand the least deviation of some observable variables from their desired values. This problem is not devoid of physical sense although the entropy of an adiabatically isolated system as a whole cannot be decreased by any external field, as is well-known.

As follows from this short preliminary discussion, the basic concepts of the theory of optimal control can naturally be employed in the quantum case. The difficulty of the problems in the quantum theory of control, as compared with the classical theory, consists in the drastic qualitative change in the complexity of describing the dynamic systems, which requires new notions and novel complicated mathematical apparatus. Operators are used in quantum mechanics instead of ordinary variables. State functions are complex-valued. There are partial differential equations or other type of equations in functional spaces instead of ordinary differential equations. Exceptionally great difficulties are encountered in problems of controlling quantum systems with macroscopic feedbacks because the description of the measurement processes requires two-level or multilevel models of the dynamical systems. Control functions often enter non-linearly in the coefficients of the equations.

However, there are no fundamental difficulties in the analysis and synthesis of systems where feedback is carried out on the microscopic, that is, quantum level, if the controller is consistently viewed as the varying part of the whole system described by an extended Hamiltonian.

At present, the quantum theory of control is in its infancy. Therefore it is only natural that we should start considering the basic principles and simplest examples allowing us to analyze the specifics of quantum system control problems. Pure ensembles are most suitable for this because mixed ones are to a certain degree intermediate between pure quantum ensembles and statistical ones.

1.3. Problems of Control in Different Descriptions

A problem in the quantum theory of systems control can be formulated in a rather general form as follows. Assume $\mathcal{V}$ is a set of admissible controls u. The mathematical nature of the set $\mathcal{V}$ can be very diverse: a set of functions, operators, vectors (including infinite-dimensional ones), real numbers, etc. The elements of this set describe, for example, such physical entities as external electromagnetic fields or fields of some other nature which are the carriers of the control. Among other things, the initial state of the controlled system, for example, $\psi(x, t_0) = u(x)$ may serve as a control. These controls may be produced by some external independent source. On the other hand, when $\mathcal{V}$ is a set of operators, it can describe a set of controls exerted on the given object by another system of the micro- or macro-level. Control in this case is carried out by selected subsystems rendered interactive, so that the system as a whole could gain some desired properties. For instance, control may consist of a Hamiltonian H_0 of the initial system supplemented by a Hamiltonian H_1 of another system so that the joint system with the Hamiltonian $H_0 + H_1$ possesses a desired property.

Generally, the objective of control is to ensure that a quantum-mechanical system possesses certain desired properties, and this can be described in the following manner. Suppose $\psi(u)$ is an element of a set and describes the state obtained by the control u by virtue of the quantum-mechanical equations of motion (for example, the Schrödinger differential equations, Heisenberg operator equations, Born integral equation, etc.). There is an operator B defined on the set of states and controls $u \in \mathcal{V}$. A subset β is given in the range of values of B. Then the objective of control can be formulated as follows: find $u \in \mathcal{V}$ such that

$$B[\psi(u), u] \in \beta. \tag{1}$$

If the condition (1) holds for some $u \in \mathcal{V}$, the system is said to be *controllable with respect to the properties given by B and β*. If (1) does not hold uniquely, we can demand that the functional $F[\psi(u), u]$ generally depending both on the state and control, should assume an extremal value.

A frequently occurring case of (1) is a finite or infinite system of equations

$$F_k[\psi(u), u] = \alpha_k, \ k = 1, 2, \ldots, \tag{2}$$

where F_k are certain functionals of $\psi(u)$ and u while α_k are given numbers. In particular, equations (2) define requirements imposed on the momenta of the distribution functions.

Numerous quantum system control problems can be cited making this general set-up more concrete. This will be done in the subsequent sections of this chapter.

1.4. Obtaining a Prescribed Pure State or a State in its Vicinity

Suppose that a system is defined by a Hamiltonian $H(q, \partial/\partial q, t, u)$, where u is control from a certain set $\mathcal{V}$. Suppose at the initial instant $t = t_0$ or before it, at $t < t_0$, the system is in the state defined by the function $\psi_0(q, t)$ (or $\psi_0(q)$). It is required to select a control $u \in \mathcal{V}$ (for $t > t_0$) such that after $t = T$ (or at $t = T$) the system would be in a desired state

$$\psi(q, t) = \psi_1(q, t), \ t > T,$$

or

$$\psi(q, t) = \psi_1(q), \ t = T.$$

Naturally one can demand at the same time that a functional with a definite meaning (for example, energy or transition time) reach an extremum.

If these equalities are infeasible or unnecessary for some reasons, one can demand that some functional of the current state $\psi(q,t)$ and of the desired state $\psi_1(q,t)$ should reach a minimum. In particular, this can be the transition of the system, due to some control (perturbation), from one stationary state into a desired stationary state or the minimization of some measure of the distance between these states.

1.5. Control with the Aim of Obtaining a Specified Probability of a Given Pure State

Suppose that the system is in a state $\psi(q,t_0)$ at the instant $t = t_0$. It is required to find a control $u \in \mathcal{V}$ such that at the instant $T > t_0$ the probability that the system is in a desired state $\psi_1(q)$ is maximal.

As is well known (Feynman and Hibbs, 1965), the amplitude of this probability is given by

$$a(T) = \int \psi_1^*(q)\psi(q,T)dq, \qquad (1)$$

and therefore it is required to find $u \in \mathcal{V}$ such that the functional

$$|a(T)|^2 = a^*(T)a(T) \qquad (2)$$

would reach either the prescribed or its maximum (or minimum) value.

One of the possible techniques for solving this kind of problem consists in the following. Suppose, for instance, that the unperturbed (uncontrolled) part of the system in the coordinate representation is defined by the Hamiltonian $H_0(q,\partial/\partial q,t)$ and the controlled part is defined by the sum $H_0(q,\partial/\partial q,t) + H_1(q,\partial/\partial q,t,u)$ where u is a control ($u \in \mathcal{V}$). Then (see Appendix 1) the state $\psi(q,t)$ obeys the integral equation

$$\psi(q,t) = \int_{t_0}^{t} \int G_0(q,t,\xi,\tau)H_1(\xi,\partial/\partial\xi,\tau,u)\psi(\xi,\tau)d\xi d\tau +$$
$$+ \int G_0(q,t,\xi,t_0)\psi_0(\xi)d\xi, \qquad (3)$$

where G_0 is the Green's function corresponding to H_0 and $\psi_0(q)$ is the initial state at $t = t_0$. The maximized functional of the probability of obtaining the desired state $\psi_1(q)$ at $t = T$ is

$$F = \left| \int \psi_1^*(q)\psi(q,T)dq \right|^2. \tag{4}$$

This problem is completely within the pattern of the optimization problems for distributed systems. In order to solve it, one can use the maximum principle for systems described by an integral equation (Butkovskiy, 1965). In the Born approximation (Davydov, 1973), the integral equation becomes an explicit integral relation between the controlled function $\psi(q,t)$ and the control u. The equations for the sought optimal control will be considerably simpler in this case (Butkovskiy, 1965, p.99).

If control of the system enters into an additive perturbation of the Hamiltonian, then the most natural representation of the states is the interaction representation. The latter is often an initial one for the investigation of many quantum phenomena and, in particular, transition processes. Here, however, the notion of a transition process has a specific meaning, somewhat deviating from the classical meaning, in particular the one usually implied in the theory of control of deterministic systems. Indeed, quantum theory does not try to describe the course of transition processes in time as a continuous and determined sequence of certain states. This notion would rather imply what is implied by transition processes in the theory of controllable stochastic Markov processes (Belavkin, 1980; Pugachev, 1957; Pugachev et al., 1974; Fleming and Richel, 1975). In fact the mathematical descriptions deal with the time variation of complex amplitudes of probabilities of the values of physical variables rather than with time variation of the physical variables themselves.

The equations for the probability amplitudes of the system's transition from one state into another are well known (Feynman et al., 1963; Feynman and Hibbs, 1965). Assume that the Hamiltonian of the whole system equals the sum $\widehat{H}_0 + \widehat{H}_1(t, u(t))$ where $\widehat{H}_0$ does not explicitly depend on time t and $\widehat{H}_1$ depends on both time and control and is finite over $[t_0, t_1]$, that is, $\widehat{H}_1(t), u(t)) \equiv 0$ at $t \notin [t_0, t_1]$. Now we can discuss a quantum transition from a stationary state into another stationary state at $t > t_1$. Suppose that E_n, ψ_n are non-multiple eigenvalues and eigenstates of the operator $\widehat{H}_0$. At the initial instant $t = t_0$, the state $\psi(q, t_0)$ can be presented as an expansion in terms of the eigenfunctions of the operator $\widehat{H}_0$:

$$\psi(q, t_0) = \sum_k a_{0k} \psi_k(q) \exp\left[-\frac{i}{\hbar} E_k t_0 \right]. \tag{5}$$

Then the amplitudes of the kth eigenstate ψ_k at an instant t obey the system of linear differential equations with controlled variable coefficients

$$i\hbar\dot{a}_k(t) = \sum_m g_{km}(t, u(t)) \exp\left[\frac{i}{\hbar}(E_k - E_m)t\right] a_m(t), \qquad (6)$$

where

$$g_{km}(t, u(t)) = (\psi_k^*, \widehat{H}_1(t, u(t))\psi_m) \qquad (7)$$

is a controlled matrix element of the perturbation Hamiltonian $\widehat{H}_1$ (see Appendix 1). At $t > t_1$, these amplitudes are constant and equal to $a_k(t_1)$. The initial condition has the form

$$a_k(t_0) = a_{0k}. \qquad (8)$$

A problem of finite control can be formulated for this equation (Feldbaum, 1971; see also Appendix 1): find a control $u(t)$ over $t_0 \leq t \leq t_1$ such that the system goes from the initial state (8) to the final desired state

$$a_k(t_1) = a_{*k}.$$

The optimization problem for this equation can be formulated as follows. Find $u(t)$ over $t_0 \leq t \leq t_1$ such that the absolute value of the scalar product $|(c, a_k(t_1))|$, where c is a non-zero vector, reaches its maximum, the initial condition being (8). Additional constraints can be imposed on the coordinates $a_k(t_1)$. For instance, some of them must equal given numbers a_{*k}.

The first (Born) approximation $a_k^1(t)$, where the initial value a_{0m} is substituted into the left hand side of equations (8) instead of $a_m(t)$, can be found explicitly:

$$a_k^1(t) = \frac{1}{i\hbar} \sum_m \int_{t_0}^t a_{0m}(\tau, u(\tau)) \exp\left[\frac{i}{\hbar}(E_k - E_m)\tau\right] d\tau. \qquad (10)$$

Now substituting $a_k^1(t)$ into the right hand side of (6) and integrating, we can find explicitly the second approximation $a_k^2(t)$, etc.

Under the appropriate conditions, this technique allows us to obtain sufficiently accurate solutions. At each step of the approximation we have explicit expressions for the amplitudes $a_k(t)$ as integrals of functions depending on the control $u(t)$. In order to optimize this kind of system, one can use the procedure of the maximum principle for integral relations (Butkovskiy, 1965, p.99).

For instance, the first approximation of the amplitude of transition from an eigenstate ψ_m to an eigenstate ψ_k is expressed by the very simple formula

$$a_k^1(t_1) = \frac{1}{i\hbar} \int_{t_0}^{t_1} (\psi_k^*, \widehat{H}_1(\tau_1, u(\tau))\psi_m) \exp\left[\frac{i}{\hbar}(E_k - E_m)\tau\right] d\tau. \qquad (11)$$

The first approximation P_k^1 of the probability of the same transition is

$$P_k^1 = |a_k^1(t_1)|^2 = \frac{1}{\hbar}|W_{km}(\omega_{mk})|^2, \qquad (12)$$

where $W_{km}(\omega)$ is the Fourier transform of the matrix element $g_{km}(t, u(t))$ and $\omega_{mk} = (E_k - E_m)/\hbar$ is the Bohr frequency of transition from the energy level E_k to the level E_m.

1.6. Obtaining the Maximum (or Minimum) Probability of a Specified Value of a Physical Quantity

Suppose we are interested in obtaining such a pure state $\psi(q, T)$ where a given physical variable A assumes a specified value α, the probability of which should be maximum. Let $\psi_\alpha(q)$ be the eigenfunction of the operator $\widehat{A}$ of A corresponding to the eigenvalue α. It is evident then that the probability that A will become α is maximum when the probability that the system is in the state $\psi_\alpha(q)$ is maximum. Therefore it is necessary (as in §1.5) to find a control $u \in V$ such that the system will occur in the state $\psi(q, T)$ that maximizes (or minimizes) the functional

$$|a(T)|^2 = a^*(T)a(T), \qquad (1)$$

where

$$a(T) + \int \psi_\alpha^*(q)\psi(q, T)dq. \qquad (2)$$

Let us point out again that the solution of this kind of problem may be found by using the techniques of optimization for systems defined by integral equations (Butkovskiy, 1965).

1.7. Obtaining a Desired Distribution of Probability Amplitudes for Values of Given Physical Quantities

This problem can be posed as follows. Assume that Λ_1 is a given part (subset) of the spectrum Λ of a physical quantity A. Suppose that $a_*(\lambda)$ is a given distribution of probability amplitudes on the given part Λ_1 of the spectrum Λ, that is, $\lambda \in \Lambda_1 \subset \Lambda$. Find a control $u(t) \in \mathcal{V}$, $0 \le t \le T$, such that the state of the system satisfies the equality

$$a_*(\lambda) = \int \psi_\lambda^*(q)\psi(q,T)dq, \ \lambda \in \Lambda_1 \subset \Lambda, \tag{1}$$

where $\psi_\lambda(q)$ are eigenfunctions corresponding to eigenvalues λ. These equalities (a system of integral equations) can be regarded as a moment problem which is generally non-linear with respect to the control $u \in \mathcal{V}$ sought.

Note that the amplitudes $a_*(\lambda)$ can be replaced by probability distributions, that is, numbers $P_*(\lambda\| = |a_*(\lambda)|^2$. It is therefore required to satisfy the system

$$P_*(\lambda) = \left| \int \psi_\lambda^*(q)\psi(q,T)dq \right|^2. \tag{2}$$

instead of the system (1).

If the validity of equations (1), (2) is for some reason infeasible or unnecessary, one can pose a problem of finding a control $u \in \mathcal{V}$ such that the current distribution $a(\lambda)$ or $P(\lambda)$ has, in a sense, a minimum deviation from the desired distribution $a_*(\lambda)$ or $P_*(\lambda)$ respectively. For instance, such a problem was prepared for solution in Feynman and Hibbs (1965), where explicit expressions were deduced for the functionals of probabilities that a harmonic oscillator would go from one state to another depending on the force acting on the oscillator. Naturally, this kind of problem can be formulated in terms of controlled statistical processes.

We can similarly pose a problem of controlling the distribution of probability amplitudes for possible values of several physical quantities. Suppose that there are several physical quantities $A_1, \ldots, A_n$. In a given state ψ each of these variables possesses a distribution of probability amplitudes $a_j(\lambda_j)$ on its spectrum of possible numbers $j = 1, \ldots, n$. It is clear that these distribution functions depend on controls applied to the system, so one can pose problems of obtaining distributions $a_j(\lambda_j)$, $j = 1, \ldots, n$, that coincide with or, in some sense, deviate least from the given distributions $a_{*j}(\lambda_j)$, $j = 1, \ldots, n$.

However, there can be fundamental physical constraints impeding the achievement of desired distributions besides the constraints imposed by the system itself. Many distributions fail to be obtained with a desired accuracy. This follows from the Heisenberg uncertainty relation (Appendix 1). In this connection, an interesting problem is the study of the influence of control on the range of the Heisenberg uncertainty and find out which form of control (consistent with the uncertainty relation) minimizes this uncertainty allowing us to obtain coherent states.

1.8. Control of Quantum Averages and Moments of Physical Quantities

Typical controlled functionals are the quantum averages $\bar{A}_1, \ldots, \bar{A}_n$ of the respective physical quantities (observables) $A_1, \ldots, A_n$. For discrete microsystems (microparticles), these can be coordinates, momenta, angular momenta, energy, etc. The averages $\bar{A}_1(t), \ldots, \bar{A}_n(t)$ of pure states, being time functions, obey the Ehrenfest equations (see Appendix 1). The averages of physical variables in a mixed state can be obtained using Equation (37) of Appendix 1. They are expressed in terms of the density operator ((36) in Appendix 1) which in turn obeys the Liouville quantum equation (Appendix 1). Therefore it is possible to pose such problems of controlling the averages that are quite similar to a whole set of problems considered in the classical theory of control: those of optimization, controllability, observability, identification, finite controllability, stability, filtration, etc. Moreover, it is easy to show that in the case when the Hamiltonian of the system is a quadratic form in the canonical variables q_k and p_k, the Ehrenfest equations for the averages $\bar{q}_k$ and $\bar{p}_k$ *coincide exactly* with the classical linear equation of motion and therefore, in order to solve the problems of control of this kind of quantum system, we can rely completely on the whole apparatus and ready solutions obtained in the theory of control of lumped parameter systems (Andreev, 1976); Krasovskii and Pospelov, 1962; N. Krasovskii, 1968; Gabasov and Kirillova 1971; Roitenberg, 1971; Zubov, 1975; A. Krasovskii, 1968; Krasovskii and Pospelov, 1962; Pugachev, 1957; Pugachev et al., 1974; Feldbaum, 1971; Feldbaum and Butkovskiy, 1971; Pontryagin et al., 1963; Voronov, 1966, 1079).

For quantum fields, the averages $\bar{A}_1(q, t), \ldots, \bar{A}_n(q, t)$ of physical quantities are functions of space coordinates, the functions describing the states of distributed parameter entities. This is why we can use here the results of the theory of control of distributed parameter systems (Butkovskiy, 1965, 1975, 1977).

The problem of control of averages can naturally be extended to involve both the first moments and other moments (for example, variance) as well. However, when the second central moments (variances) are involved, it is essential to take into account the fundamental constraint imposed by the Heisenberg uncertainty relation ((14a)–(14c), Appendix 1), as was indicated in the preceding section. Therefore, the formulation of problems of control of quantum systems requires a thorough analysis of their mathematical and physical consistency.

Let us remark that apart from the kth moments the formulations of a number of control problems may include (as controlled functionals) other functions of the moments, for example, trigonometric ones or functions of a special kind, as well as non-linear functionals (Butkovskiy and Pustylnikov, 1980).

1.9. Control of the Distributions of Eigenvalues of Physical Quantities

An important practical problem is that of the control of the spectrum of eigenvalues of an operator $\widehat{A}$. Indeed, suppose that $\widehat{A}$ depends in a certain way on a function $u(q)$ that can be arbitrarily selected from a class of admissible functions $u(q) \in \mathcal{V}_q$, that is, this is a case of static control that does not depend on t. For instance $u(q)$ can be a potential. Then it is possible to pose the problem of finding a control $u(q) \in \mathcal{V}_q$ such that the operator $\widehat{A}$ have a desired spectral function $E_*(\lambda)$ or such that a certain measure of the deviation of the spectral function $E(\lambda)$ corresponding to $u(q)$ from the desired function $E_*(\lambda)$ be a minimum.

We can also pose problems of control of functionals of the spectral function $E_*(\lambda)$, for example, to obtain an extremal (or desired) value of the spectral radius. This is also the case of posing the problem of obtaining eigenfunctions of some desired form or those of a similar form. An important particular case is the problem of the energy spectrum, where $\widehat{A} \equiv \widehat{H}$ is the Hamiltonian of the quantum system.

Similar problems appear in the theory of elastic systems (membranes, shells, etc.), where it is necessary to find, for instance, the density distribution in a body such that the eigenfrequencies are in the vicinity of the desired ones. These problems are also essential in control theory (Lurie, 1975).

It is interesting to note that control of a spectrum can be carried out, in a sense, by means of controls depending on time, as it was shown in Vinogradskaya and Girko (1980) and in Girko and Vinogradskaya (1979). It was pointed out in Waugh (1976) that an appropriate time-dependent control (a sequence of certain impulses of the magnetic field in this case) can make an 'effective' Hamiltonian possess better spectral properties as compared with the original one.

1.10. Control of Operators of Physical Quantities

The problem discussed in the preceding section is a particular case of more general problems where arbitrary controlled operators are considered. The objective of control in them can also be to obtain some desired or extremal operator properties expressed in some terms. The preceding section discussed a Hamiltonian where, for instance, a potential $u(q)$ is to be found, the desired properties being expressed in terms of a spectral function $E(\lambda)$ and the extremal ones in terms of the spectral radius.

The Heisenberg equations ((7), (8), (11) in Appendix 1) afford a more general approach to control. These equations allow us to pose the problem of control of the operators of physical quantities directly, and this leads to a great variety of control problems.

Besides the above-mentioned examples of problems of obtaining desired spectral properties of operators, one can cite a number of problems of obtaining desired functional properties of controlled operators of physical quantities. As was noted above, the control u itself can be an operator $\hat{u}$ from a given set of operators $\widehat{\mathcal{V}}$. For instance, we can pose the problem of finding a control $\hat{u} \subset \widehat{\mathcal{V}}$ such that the commutator $[\widehat{A}, \widehat{B}]$ of two controlled operators $\widehat{A}(u), \widehat{B}(u)$ is equal to zero identically or equal to zero at certain values of the independent variables. It is clear that this is not always possible to demand because of the Heisenberg uncertainty relation. However, a general investigation in this direction is not purely academic.

For instance, one of the problems solved in Waugh (1976) concerning NMR is that of finding a control $u(t)$ (a combination of constant and impulse magnetic fields) such that the initial Hamiltonian changes in an essential way. Thus, the spin Hamiltonian becomes independent of the other components of the general

Hamiltonian in the sense that the spin Hamiltonian commutes with them. At the same time, inversion of the Hamiltonian occurs, and therefore time inversion of the internal interaction processes occurs, and other interesting phenomena take place.

It is said about these phenomena that this is a vivid example of the 'alchemy' of spin systems, that is, a change of the system's interaction Hamiltonian as the experimenter wishes. However, this 'alchemy' can be tackled quite consciously and purposefully from the general positions of the theory and methods of control of quantum systems.

1.11. Measurement in Systems with Feedback

As follows from the fundamentals of quantum mechanics (see, for instance, Neumann (1932)), measurements in a quantum system, carried out in order to find its current state, perturb the system unpredictably. This fact brought forth various problems of quantum filtration, identification, and control (Belavkin, 1980; Petrov and Uskov, 1975; Helström, 1976; Kholevo, 1980). Exceptions are special methods of non-perturbative or almost non-perturbative control and measurement. These methods were investigated by a number of researchers (Belavkin, 1975; Dodonov et al., 1980). We shall consider below the general scheme of the occurrence of the Markov process as a result of a sequence of measurements and the possibility of controlling the process.

When a quantum system is included into a loop with a macroscopic feedback, it behaves like a macroscopic block with a statistical operator. In the general case, the Hamiltonian of such a system may depend on the control $u(t)$ and statistical external disturbances $\nu(t)$:

$$\widehat{H} = \widehat{H}[u(t), \nu(t)]. \tag{1}$$

Uncertainty is due both to $\nu(t)$ and the influence of the macroscopic instrument measuring the output observable quantity A of the quantum system.

Assume that the measurement of the quantity A occurs periodically at discrete instants $t_k = k\tau$ ($k \in N$, being an integer). Suppose furthermore that $u(t)$ and $\nu(t)$ are constant between the instants of measurement. Then the evolution of the state operator $\hat{\rho}$ occurs as follows:

$$\hat{\rho}(t_{k+1}) = \int \widehat{U}(u_k, \nu_k)\hat{\rho}(t_k)\widehat{U}^{-1}(u_k, \nu_k)dW_k(\nu_k), \tag{2}$$

where $\widehat{U}(u_k, \nu_k)$ is the evolution operator over the interval at the given u_k, ν_k and $W_k(\nu_k)$ is the distribution function of the sequence of the statistical variables ν_k which we assume to be independent here. Suppose that the spectrum of the operator $\widehat{A}$ is non-degenerate. Then owing to the reduction of the statistical operator in the process of measurement carried out at an instant t_k, we have $\hat{\rho}(t_k) = |a_k\rangle\langle a_k|$, where $|a_k\rangle$ is the eigenfunction of $\widehat{A}_k$ corresponding to the measured value a_k and $\langle a_k|$ is its conjugate function. In compliance with the postulate of observables (Appendix 1), we obtain

$$P(a_{k+1} \in \Delta | a_k, u_k) = \mathrm{Sp}\left[\pi(\Delta) \int \widehat{U}(u_k, \nu_k) |a_k\rangle\langle a_k| \widehat{U}^{-1}(u_k, \nu_k) dW_k(\nu_k)\right].$$
(3)

This conditional distribution of probabilities for a_{k+1} for known a_k and u_k defines the controlled discrete Markov process. Using (3), let us find the conditional expected value and variance of the shift of the measured value $\Delta_k = a_{k+1} - a_k$ per single step. A straightforward calculation shows that

$$g(a_k, u_k) = \mathrm{E}(\Delta_k | a_k, u_k) = \iint (a_{k+1} - a_k) |U(a_{k+1}, a_k; u_k, \nu_k)|^2 da_{k+1} dW_k(\nu_k),$$

$$h(a_k, u_k) = \mathrm{E}(\Delta_k^2 | a_k, u_k) =$$
$$= \iint (a_{k+1} - a_k)^2 |U(a_{k+1}, a_k; u_k, \nu_k)|^2 da_{k+1} dW_k(\nu_k).$$

Here $U(a_{k+1}, a_k; u_k, \nu_k)$ is the kernel of the evolution operator $\widehat{U}(u_k, \nu_k)$ in the A-representation. Inasmuch as $\frac{1}{i\hbar}\widehat{H}(u_k, \nu_k)$ is the infinitesimal operator of the evolution group, we have for small τ

$$|U(a_{k+1}, a_k; u_k, \nu_k)|^2 = \delta(a_{k+1} - a_k) - \frac{\tau^2}{\hbar^2}|H(a_{k+1}, a_k; u_k, \nu_k)|^2 + O(\tau^4), \quad (4)$$

where $H(a_{k+1}, a_k; u_k, \nu_k)$ is the kernel of the energy operator in the A-representation which we assume to be given in the description of the system. Considering (4), we obtain

$$g(a_k, u_k) = -\frac{\tau^2}{\hbar^2} \iint (a_{k+1} - a_k) |H(a_{k+1}, a_k; u_k, \nu_k)|^2 da_{k+1} dW_k(\nu_k) + O(\tau^4),$$
(5)

$$h(a_k, u_k) = -\frac{\tau^2}{\hbar^2} \iint (a_{k+1} - a_k)^2 |H(a_{k+1}, a_k; u_k, \nu_k)|^2 da_{k+1} dW_k(\nu_k) + O(\tau^4).$$
(6)

The rates of drift and diffusion to within $O(\tau^3)$ equal $\frac{1}{\tau}g(a_k, u_k)$ and $\frac{1}{\tau}h(a_k, u_k)$ respectively. Using these rates, we can arrive at the Fokker-Planck equation

$$\frac{df(\theta, a)}{d\theta} + \frac{\partial}{\partial a}[g(a, u)f(\theta, a)] = \frac{1}{2}\frac{\partial^2}{\partial a^2}[h(a, u)f(\theta, a)], \qquad (7)$$

describing the evolution of the probability density $f(\theta, a)$ observed in dimensionless time $\theta = t/\tau$ in the presence of control. Equation (7) takes into account the influence of measurements carried out periodically after each interval τ.

It is easy to see that in real time t the evolution rate is proportional to τ and therefore becomes infinitesimal as $\tau \to 0$. In other words, continuous measurement would result in rendering the observable constant for the whole period of measurement. The evolution only resumes after the macroscopic instrument is disconnected from the quantum system. When the instrument is connected to it, a drastic 'loss of sensitivity' of the quantum system occurs, and it does not respond to the changing microsituation ν_k, u_k.

At constant control and constant perturbation, the function $P(b|a, u, \nu) = |U(b, a; u, \nu)|^2$ is the conditional probability density of observing a value b of the observable A at the end of an interval τ when a value a was observed at its beginning. Applying the Bayes formula, we can find the conditional probability density for ν_k if the results of measurement a_{k+1}, a_k and control u_k are known:

$$P_k(\nu_k | a_{k+1}, a_k, u_k) = \frac{|U(a_{k+1}, a_k; u_k, \nu_k)|^2 W_k'(\nu_k)}{\int |U(a_{k+1}, a_k; u_k, \nu_k)|^2 W_k'(\nu_k)d\nu_k}. \qquad (8)$$

It is assumed here that the distribution density is $W_k'(\nu_k)$. Equation (8) makes it possible to evaluate the perturbation, for example, in a gravitation wave experiment with a quantum-mechanical test body.

From the viewpoint of the quantum theory of measurement, we can pose a problem: find an estimate ν_k^* of ν_k such that the conditional average risk of decision-making attains its minimum:

$$\mu(a_{k+1}, a_k, u_k) = \min_{\nu_k} \mathbf{E}\{r(\nu_k, u_k)|a_{k+1}, a_k, u_k\}.$$

The control problem consists in minimizing the average risk of a random realization of a_{k+1} when a_k is given: it is necessary to find

$$u_k(a_k) = \arg\min_{a_k} \int \mu(a_{k+1}, a_k, u_k) \int |U(a_{k+1}, a_k; u_k, \nu_k)|^2 dW_k(\nu_k)da_{k+1}. \qquad (9)$$

This problem becomes essentially more complicated if the statistical relation between the ν_k is taken into account. It is a problem of dual control (Feldbaum, 1971) requiring consistency of probabilistic measures in the controlled statistical process (Helström, 1976; Kholevo, 1980).

CHAPTER 2

Controllability and Finite Control of Quantum Processes (Analytical Methods)

2.1. Control of Pure States of Quantum Processes

General problems of control of quantum-mechanical systems and approaches to their solution were discussed in the preceding chapter. We shall describe below the problem of controllability and the problems of control of pure states and averages (Butkovskiy and Samoilenko, 1979a, 1979b, 1980).

Let us consider a quantum-mechanical system whose Hamiltonian has the form

$$\widehat{H}_t = \widehat{H}^0 - \sum_{j=1}^{n} \widehat{B}^j u_j(t), \tag{1}$$

where $\widehat{H}^0$ is the unperturbed stationary part (that is, not depending on time explicitly) of the complete energy operator $\widehat{H}_t$, the $u_j(t)$ are controls (which are components of the macroscopic force field) and the $\widehat{B}_j$ are their corresponding energy operators. Assume that

$$u_j(t) \equiv 0 \quad \text{for } t < 0 \text{ and } t > T, \tag{2}$$

and suppose that the results of measurements before the start of control are such that the system has a certain value of energy E_m and certain values $A^i_{k_i}$ of a number of other physical quantities $\widehat{A}^i$, $i = 1, \ldots, N$ which including the energy, comprise a complete set of observables describing a pure stationary state with vector-function

$$\psi_{m, k_1, \ldots, k_N}(t) = \exp\left(-\frac{iE_m t}{\hbar}\right) \psi_{m, k_1, \ldots, k_N}. \tag{3}$$

As is well known from the quantum theory (Blokhintsev, 1981a; Davydov, 1973; Dirac, 1958; Landau and Lifshits, 1963) (see also Appendix 1), certain values of the observables belonging to the complete set, are the eigenvalues of the corresponding operators $\widehat{H}^0, \widehat{A}^1, \dots, \widehat{A}^N$ and $\psi_{m,k_1,\dots,k_N}$ are the orthonormal eigenvectors common for all these operators, so that

$$
\blacksquare \quad
\begin{aligned}
\widehat{H}^0 \psi_{m,k_1,\dots,k_N} &= E_m \psi_{m,k_1,\dots,k_N}, \\
\widehat{A}^1 \psi_{m,k_1,\dots,k_N} &= A^1_{k_1} \psi_{m,k_1,\dots,k_N}, \\
\widehat{A}^N \psi_{m,k_1,\dots,k_N} &= A^N_k \psi_{m,k_1,\dots,k_N}, \\
(\psi_{m,k_1,\dots,k_N}, \psi_{n,l_1,\dots,l_N}) &= \delta_{mn}, \; \delta_{k_1 l_1}, \dots, \delta_{k_N l_N}, \quad \square
\end{aligned}
\tag{4}
$$

where

$$
\delta_{mn} = \begin{cases} 1 & \text{for } m = n, \\ 0 & \text{for } m \neq n \end{cases}
\tag{5}
$$

is the Kronecker symbol. The spectra may be both continuous and discrete. For definiteness, we can assume that they are discrete, that is, the system performs finite motion (Landau and Lifshits, 1963) or, as they say, the system is in a bound state. However, it would have been sufficient to assume the separability of the state space.

The indices $\mu = \{m, \mathbf{k}\} = \{m, k_1, \dots, k_N\}$ of the states are multicomponent or vectorial (they are multi-indices). For the moment, the representation is not made specific. Generally, stationary energy levels are degenerate; the system is defined in this case by the introduction of the indices of other values of the observables from the complete set. Each set of values of the observables corresponds to one, and only one, state vector-function.

Let us formulate the objective: to transfer the system in the interval $0 \leq t \leq T$ from an initial state μ to another state ν by means of controls $u_j(t)$, $j = 1, \dots, n$, while at the same time minimizing the functional of average risk chosen here as the sum of quadratic forms of the state vector-function and controls:

$$
R = \int_0^T \left[\left(\psi(t), \widehat{R}_t \psi(t) \right) + \sum_{i,j=1}^n r_{ij}(t) u_i(t) u_j(t) \right] dt,
\tag{6}
$$

where $\widehat{R}_t$ is the risk operator, which may depend on time explicitly, and $r_{ij}(t)$ are the coefficients of the cost of control. Without limiting ourselves by a concrete basis, let us comment on this problem in various representations of transition from state to state.

The Schrödinger representation. Schrödinger's equation of motion (Appendix 1) for the vector ψ has the form

$$i\hbar\frac{d\psi(t)}{dt} = \widehat{H}^0\psi(t) - \sum_{j=1}^{n} u_j(t)\widehat{B}^j\,\psi(t). \tag{7}$$

The solution must satisfy the boundary conditions

$$\psi(0) = \psi^0_{m,k_1,\dots,k_N}, \psi(T) = \exp\left(-\frac{iE_nT}{\hbar}\right)\psi^1_{n,l_1,\dots,l_N} \tag{8}$$

and provide the minimum for the functional (6) in the class of control functions $u_j(t)$ that are square integrable on the interval $[0,T]$. Owing to the hermiticity of the energy operator, the norm of the solution of equation (7) equals unity and is automatically constant for all time, therefore the control is carried out on the unit sphere in Hilbert space. The phase factor in the vector-function is inessential in many cases because it does not influence the quantum-mechanical averages and therefore may be omitted. This is why instead of (8) we can write

$$\psi(0) = \psi^0_\mu = \psi^0_{m,k_1,\dots,k_N}, \quad \psi(T) = \psi^1_\nu = \psi^1_{n,l_1,\dots,l_N}. \tag{9}$$

If the Schrödinger representation is chosen, the form of the functional (6) does not vary with time.

The Heisenberg representation. Since stationary states can be completely defined by the simultaneously measurable quantities $E_m, A^1_{k_1}, \dots, A^N_{k_N}$ of the complete set, these quantities can be prescribed instead of the initial and final state vectors. Transition of the system from one stationary state into another is equivalent to the replacement of some eigenvalues of the operators of the complete set by other eigenvalues:

$$E_m \to E_{P_0(m)}, \ A^1_{k_1} \to A^1_{P_1(k_1)}, \dots, \ A^N_{k_N} \to A^N_{P_N(k_N)}$$

by permuting the $P_0(m), P_1(k_1), \dots, P_N(k_N)$. Using the projection operators $\widehat{\Pi}^0_m, \widehat{\Pi}^1_{k_1}, \dots, \widehat{\Pi}^N_{k_N}$ onto the subspaces corresponding to the eigenvalues $E_m, A^1_{k_1}, \dots, A^N_{k_N}$, the operators of the complete set before the start of control can be represented as

$$\widehat{H}^0(0) = \sum_m E_m\widehat{\Pi}^0_m, \quad \widehat{A}^1(0) = \sum_{k_1} A^1_{k_1}\widehat{\Pi}^1_{k_1}, \dots,$$
$$\widehat{A}^N(0) = \sum_{k_N} \widehat{A}^N_{k_N}\widehat{\Pi}^N_{k_N}. \tag{10}$$

As a result of control, one of the eigenvalues (the initial one) of each of the operators is replaced by another (the final one). Therefore at the final instant of the control interval the operators of the complete set must become:

$$\blacksquare \quad \widehat{H}^0(T) = \sum_m E_{P_0(m)} \widehat{\Pi}_m^0 = \sum_m E_m \widehat{\Pi}_{P_0^{-1}(m)}^0,$$

$$\widehat{A}^1(T) = \sum A_{P_1(k_1)}^1 \widehat{\Pi}_{k_1}^1 = \sum A_{k_1}^1 \widehat{\Pi}_{P_1^{-1}(k_1)}^1, \tag{11}$$

$$\cdot \quad \cdot \quad \cdot \quad \cdot \quad \cdot \quad \cdot \quad \cdot \quad \cdot \quad \cdot \quad \cdot$$

$$\widehat{A}^N(T) = \sum A_{P_N(k_N)}^N \widehat{\Pi}_{k_N}^N = \sum A_{k_N}^N \widehat{\Pi}_{P_N^{-1}(k_N)}. \quad \square$$

Taking into account the fact that every operator of the complete set of the simultaneously measurable quantities commutes with the unperturbed Hamiltonian, we can write

$$\left[\widehat{H}^0(t), \widehat{H}_t(t) \right] = -\sum_{j=1}^{n_1} \left[\widehat{H}^0, \widehat{B}^j(t) \right] u_j(t),$$

$$\left[\widehat{A}^j(t), \widehat{H}_t(t) \right] = -\sum_{j=1}^{n_1} \left[\widehat{A}^j(t), \widehat{B}^j(t) \right] u_j(t). \tag{12}$$

Hence equation (7) becomes homogeneous and bilinear

$$i\hbar \frac{d\widehat{H}^0(t)}{dt} = -\sum_{j=1}^{n_1} \left[\widehat{H}^0(t), \widehat{B}^j(t) \right] u_j(t),$$

$$i\hbar \frac{d\widehat{A}^i}{dt} = -\sum_{j=1}^{n_1} \left[\widehat{A}^i(t), \widehat{B}^j(t) \right] u_j(t), \tag{13}$$

$$i = 1, 2, \ldots, N.$$

These must be supplemented by the evolution equation of the risk operator and the operators $\widehat{B}^k$:

$$\frac{d\widehat{R}_t(t)}{dt} = \frac{\partial \widehat{R}_t(t)}{\partial t} + \frac{1}{i\hbar} \left[[\widehat{R}_t(t), \widehat{H}^0(t)] - \sum_{j=1}^{n_1} [\widehat{R}_t(t), \widehat{B}^j(t)] u_j(t) \right],$$

$$i\hbar \frac{d\widehat{B}^k(t)}{dt} = [\widehat{B}^k(t), \widehat{H}^0(t)] - \sum_{j=1}^{n_1} [\widehat{B}^k(t), \widehat{B}^j(t)] u_j(t). \tag{14}$$

The expression for the average risk takes on the form

$$R = \int_0^T \left[\left(\psi, \widehat{R}_t(t)\psi \right) + \sum_{i,j=1}^{n_1} r_{ij}(t) u_i(t) u_j(t) \right] dt, \tag{15}$$

where quantum-mechanical averaging is carried out by means of the initial
state vector ψ which does not vary in the process of control. A necessary
condition of controllability follows directly from equations (13): *each of the
operators $\widehat{H}^0, \widehat{A}^1, \ldots, \widehat{A}^N$ must not commute with at least one operator $\widehat{B}^j$, $j =
1, \ldots, n_1$*. This shows the effectiveness of algebraic methods of investigating
controllability; the method will be discussed in more detail and generalized
below.

The interaction representation. This representation is intermediate between the first two (Appendix 1). Here the operators vary in an uncontrollable
way in compliance with equation (A1.14), [1] where the unperturbed Hamiltonian $\widehat{H}^0$ from (A1.12) must be used. Therefore the operator $\widehat{H}^0$ itself and
all the operators of the complete set commuting with $\widehat{H}^0$ are not subject to
evolution. Since they do not depend on time explicitly, they remain constant
in the process of control. The risk operator and the operators $\widehat{B}^j$, if they do
not commute with $\widehat{H}^0$, vary in accordance with the equations

$$i\hbar\frac{d\widehat{R}_t(t)}{dt} = \left[\widehat{R}_t(t), \widehat{H}^0\right], i\hbar\frac{d\widehat{B}^j(t)}{dt} = \left[\widehat{B}^j(t), \widehat{H}^0\right], \; j = 1, \ldots, n_1. \qquad (16)$$

whose solutions in view of (5) are

$$\widehat{R}_t(t) = \exp\left(\frac{i}{\hbar}\widehat{H}^0 t\right)\widehat{R}_t \exp\left(-\frac{i}{\hbar}\widehat{H}^0 t\right), \qquad (17)$$

$$\widehat{B}^j(t) = \exp\left(\frac{i}{\hbar}\widehat{H}^0 t\right)\widehat{B}^j \exp\left(-\frac{i}{\hbar}\widehat{H}^0 t\right), \qquad (18)$$

Any change of the state vector in the interaction representation occurs
between the values

$$\psi(0) = \psi^0_{m,k_1,\ldots,k_N}, \; \psi(T) = \psi^1_{n,l_1,\ldots,l_N} \qquad (19)$$

in keeping with the equation

$$i\hbar\frac{d\psi}{dt} = -\sum_{j=1}^{n_1} u_j(t)\exp\left(\frac{i}{\hbar}\widehat{H}^0 t\right)\widehat{B}^j \exp\left(-\frac{i}{\hbar}\widehat{H}^0 t\right)\psi(t), \qquad (20)$$

which in the energy representation is simpler than (7). The risk operator and
the state vector, both varying with time, must be substituted into the functional

---

[1] (A1.12) refers to formula (12) of Appendix 1. All further references to the
Appendices will be written similarly.

(6). In most cases, owing to the simplicity of the unperturbed Hamiltonian, we can succeed in finding a sufficiently simple, usually exponential expression for the corresponding evolution operator.

We draw attention to the fact that the mathematical form of the problem varies considerably with different representations of state and in a number of cases a suitable choice of the representation makes the solution easier. However, we cannot decide between these various forms without making the Hamiltonian and the other operators of the system more specific.

In the analysis of controllability in a small neighbourhood of the initial state, the Schrödinger representation is often preferable because it does not require any preliminary transformations and has a more intuitive geometrical interpretation. However, if the deviation from the initial state is large, it becomes essential that controls enter Schrödinger's equation multiplicatively and therefore, instead of a linear problem, we face a parametric one which hinders the application of the general criteria of controllability.

On the other hand, the Heisenberg representation, even in its general formulation, has led to the derivation of the most essential necessary conditions of controllability in the entire state space. This idea rests on the symmetry properties of quantum systems and can give rise to a refinement of the controllability conditions in their algebraic form. There is the attractive possibility of applying the Heisenberg representation to equations that are Hamiltonian in the classical case and include linear controls. Here the quantum equations written in the Heisenberg representation contain linear additive controls, and this enables us to apply known criteria of controllability directly to the operators of generalized momenta and coordinates.

Finally, the interaction representation is often most convenient for both the investigation of many general problems and the solution of many concrete problems of quantum control theory. No other, more special, representation of state transition will be discussed here.

Although the set-up of the problem given above is not the most general one, it is still too complicated, and therefore one cannot obtain an explicit solution of it directly without a gradual generalization of particular results. This is why we shall not tackle the problem of minimizing the functionals at the outset. Rather we shall study the controllability of pure quantum states. It seems that many specific features of controlled quantum systems can be revealed in this way.

We conclude this section by pointing out the possibility of posing the above problem of controlling pure states in the first approximation as a moment problem (Butkovskiy, 1965; N. Krasovskii, 1968) which is generally non-linear (Butkovskiy and Dustylnikov, 1980). In fact, the eigenvector of the operator of a physical quantity A possessing a given value λ_m in the prescribed final state can be regarded as the vector of the final state. Its expansion in energy functions produces a set of moment equalities that are to be accommodated by means of controls.

2.2. Local Controllability in the Vicinity of a Pure State

As was stated above, controls $u_j(t)$ enter non-linearly into the equations of motion of the state vector and often multiplicatively. Inasmuch as the general methods of investigating complete controllability of non-linear systems are rarely constructive, we begin by appealing to a theorem on local controllability of dynamic systems controlled by means of linear approximations (N. Krasovskii, 1968). According to this theorem, if a non-linear system affords linearization in the vicinity of the initial point and the system is controllable in its linear approximation, then the non-linear system is also controllable in a small enough vicinity of the same initial point.

In quantum mechanics, an approximation where the perturbation is linear is often employed. It is called *Born's first approximation* in the theory of scattering. It can be said therefore that local controllability is the controllability in the Born approximation. This can render the solution of complicated problems essentially easier, in particular when investigating processes of control of mixed states in statistical quantum mechanics. A solution found with the aid of the Born approximation can be refined by subsequent iterations.

In considering a quantum system, one should be aware of the fact that in view of the hermiticity of the Hamiltonian, the end of its state vector moves on the surface of the unit sphere under any control. In particular, a sequence of stepwise controls can conveniently be regarded as a sequence of unitary transformations of the Hilbert space of state vectors with constant Hamiltonian.

Let us take an initial state defined by a vector ψ^0. We seek a solution of

Schrödinger's equation (1.7) [1] in a small neighbourhood of it in the form

$$\psi(t) = \psi^0(t) + \xi(t), \tag{1}$$

where $\psi^0(t)$ is the solution of the unperturbed equation

$$i\hbar\frac{d\psi^0(t)}{dt} = \widehat{H}^0\psi^0(t) \tag{2}$$

with initial condition $\psi^0(0) = \psi^0$ in the following form:

$$\psi^0(t) = \exp\left(-\frac{i}{\hbar}\widehat{H}^0 t\right)\psi^0. \tag{3}$$

Then the vector $\xi(t)$ must satisfy the zero initial condition

$$\xi(0) = 0, \tag{4}$$

and the linear equation of its first approximation can be written in the form

$$i\hbar\frac{d\xi(t)}{dt} = \widehat{H}^0\xi(t) - \sum_{j=1}^{n_1} u_j(t)\widehat{B}^j \exp\left(-\frac{i}{\hbar}\widehat{H}^0 t\right)\psi^0. \tag{5}$$

Successive local approximations will approximate the equation on the unit sphere.

In order to investigate controllability, it is best to employ a generally accepted mathematical apparatus. Let us choose a representation in some basis. Most suitable is the basis of stationary states of the unperturbed system with the orthonormalized vector $\psi_{m,k_1,\dots,k_N} = \psi_{m\mathbf{k}}$. This is, in fact, the energy representation, but if the energy levels are degenerate, then it must be supplemented by other quantities that together with the energy form a complete set. This can be called the *extended energy representation*, or $(E, \bar{A})$-*representation*. The matrix of the unperturbed energy operator is diagonal, as in the usual energy representation, and equation (5) looks simple enough:

$$i\hbar\frac{d\xi_{m\mathbf{k}}}{dt} = E_m\xi_{m\mathbf{k}} - \sum_{j=1}^{n_1} u_j(t)\sum_{n\mathbf{l}} B^j_{m\mathbf{k},n\mathbf{l}} \exp\left(-\frac{i}{\hbar}E_n t\right)\psi^0_{n\mathbf{l}}, \tag{6}$$

---

[1] Each reference to a formula from another section of the same chapter consists of the section number and the formula number. When the reference is to another chapter, the chapter number precedes the reference. Thus (1.7) refers to formula (7) in §1 of the same chapter.

where the

$$\psi_{nl}^0 = \langle nl|\psi^0\rangle, \xi_{mk} = \langle mk|\xi\rangle, \; B_{mk,nl}^j = \langle mk|\widehat{B}^j|nl\rangle$$

are the components of the vectors ψ_0, ξ and of the matrix elements $\widehat{B}^j$, $j = 1, \ldots, n$, respectively. [1]

We apply the criterion of controllability for linear non-stationary systems (Andreev, 1976; N. Krasovskii, 1968; Roitenberg, 1971) to the linear systems described by equations (6). To this end we form the so-called controllability Gramian, whose kernel for the equation

$$\frac{dx}{dt} = \widehat{A}(t)x + \widehat{G}(t)u \tag{7}$$

has the form [2]

$$\widehat{W}(t_0, t) = \int_{t_0}^t \widehat{\Phi}(t_0, \tau)\widehat{G}(\tau)\widehat{G}^+(\tau)\widehat{\Phi}^+(t_0, \tau)d\tau. \tag{8}$$

Here $\widehat{\Phi}(t, \tau) = \widehat{X}(t)\widehat{X}^+(\tau)$, where $\widehat{X}(t)$ is the fundamental matrix, or the evolution operator of the homogeneous system. If $t_0 = 0$, $t = T$, we can write

$$\widehat{W}(T) = \widehat{W}(0, T) = \int_0^T \widehat{X}^+(t)\widehat{G}(t)\widehat{G}^+(t)\widehat{X}(t)dt. \tag{9}$$

The controllability criterion is based on the fact that the range of values of the Gramian coincides with the domain of controllability. If there is a solution for the equation $\widehat{W}(T)y = \widehat{X}^+(T)x_T$, the admissible control corresponding to this solution is defined by the formula

$$u(t) = \widehat{G}^+(t)\widehat{X}(T)y. \tag{10}$$

Since the operator of the homogeneous system in the chosen representation is diagonal, we can easily find its functional matrix

$$X_{mk,nl}(t) = \delta_{mn}\delta_{kl} \exp\left(-\frac{i}{\hbar}E_m t\right).$$

This leads to the following expressions for the matrix elements of the controllability Gramian:

$$W_{mk,nl}(T) = \sum_{j=1}^{n_1} W_{mk,nl}^j(T), \tag{11}$$

---

[1] Here $\langle \cdot|\cdot\rangle$ is Dirac's bra-ket notation (see Appendix 1).

[2] Here and throughout, the superscript 'plus' denotes the Hermitian adjoint.

$$W^j_{m\mathbf{k},n\mathbf{l}}(T) = \int_0^T b^j_{m\mathbf{k}}(t)b^{j*}_{n\mathbf{l}}(t)dt, \tag{12}$$

$$b^j_{m\mathbf{k}}(t) = \frac{i}{\hbar}\sum_{m'\mathbf{k}'} B^j_{m\mathbf{k},m\mathbf{k}'}(t)\psi^0_{m'\mathbf{k}'}, \tag{13}$$

$$B^j_{m\mathbf{k},m'\mathbf{k}'}(t) = \exp\left(\frac{i}{\hbar}E_m t\right)\widehat{B}^j_{m\mathbf{k},m'\mathbf{k}'}\exp\left(-\frac{i}{\hbar}E_{m'}t\right). \tag{14}$$

Equation (11) shows that because the domain of controllability coincides with the range of the Gramian, the subspace of controllability for a given initial state can be presented as the union of the controllability subspaces corresponding to each control $u_j(t)$ separately. The expression (14) can be interpreted as a matrix element for the operator $\widehat{B}^j$ in the Heisenberg representation.

Note that the same result could be obtained more easily by using the interaction representation. Indeed, we can set $\psi(t) = \psi^0 + \xi(t)$, $\xi(t) \in \Xi$ ($\xi \perp \psi^0$), linearize (1.20) and thus obtain the following equation for $\xi(t)$ in the interaction representation:

$$\dot{\xi}(t) = \frac{i}{\hbar}\sum_{j=1}^{n_1} u_j(t)\exp\left(\frac{i}{\hbar}\widehat{H}^0 t\right)\widehat{B}^j\exp\left(-\frac{i}{\hbar}\widehat{H}^0 t\right)\psi^0.$$

It corresponds to the case where $\widehat{A}(t) = 0$ in (7). The corresponding Gramian is simpler:

$$\widehat{W}(T) = \int_0^T \widehat{G}(t)\widehat{G}^+(t)dt,$$

which in our case gives

$$\widehat{W}(T) = \sum_{j=1}^{n_1}\widehat{W}^j(T), \tag{15}$$

$$\widehat{W}^j(T) = \int_0^T |\widehat{B}^j(t)|\psi^0\rangle\langle\psi^0|\widehat{B}^j(t)|dt, \tag{16}$$

where

$$\widehat{B}^j(t) = \exp\left(\frac{i}{\hbar}\widehat{H}^0 t\right)\widehat{B}^j\exp\left(-\frac{i}{\hbar}\widehat{H}^0 t\right). \tag{17}$$

It is evident that these formulas written in operator form give the preceding formulas (11)–(14) in the extended energy representation.

Using this apparatus, we continue the investigation of controllability in the Born approximation. We can employ the functional treatment of the controllability theorem (Andreev, 1976; Roitenberg, 1971), according to which a non-stationary linear system is controllable in a controllability subspace Ξ if and only if the strict inequality

$$\langle \xi^T | \widehat{W}(T) | \xi^T \rangle > 0 \tag{18}$$

holds for some finite T for every $\xi^T = \xi(T) \in \Xi$. We can verify it using (16) and (17); we then find that controllability demands that the inequality

$$\sum_{j=1}^{n_1} \int_0^T \left| \langle \xi^T | \widehat{B}^j(t) | \psi^0 \rangle \right|^2 dt > 0 \quad \text{for all } \xi^T \in \Xi \tag{19}$$

should hold for some $j = 1, \ldots, n$. In other words, it is required that for all $\xi T \in \Xi$

$$\int_0^T |\langle \xi^T | \exp\left(\frac{i}{\hbar} \widehat{H}^0 t\right) \widehat{B}^j \exp\left(-\frac{i}{\hbar} \widehat{H}^0 t\right) | \psi^0 \rangle|^2 dt \neq 0 \tag{20}$$

for at least one $j = 1, \ldots, n_1$. To achieve this, because the integrand is continuous, *it is necessary and sufficient that for every* $\xi^T \in \Xi$ *the inequality* (21) *should hold for some* $t \in (0, T)$ *and some* $j = 1, \ldots, n_1$, *dependent on the choice of* ξ^T:

$$\langle \xi^T | \exp\left(\frac{i}{\hbar} \widehat{H}^0 t\right) \widehat{B}^j \exp\left(-\frac{i}{\hbar} \widehat{H}^0 t\right) | \psi^0 \rangle \neq 0. \tag{21}$$

This criterion agrees with the necessary condition obtained above for controllability of stationary states, namely, at least one operator B must not commute with the Hamiltonian H. For if the latter condition does not hold, then in the present instance the inequality (21) reduces to $\langle \xi^T | \widehat{B}^j | \psi^0 \rangle \neq 0$. But if $\widehat{B}^j$ commutes with $\widehat{H}^0$ and ψ^0 defines a stationary state, then $\widehat{B}^j \psi^0 = b^0 \psi^0$, where b^0 is the eigenvalue of the operator $\widehat{B}^j$ in the state ψ^0. As a result we obtain $b^0 \langle \xi^T | \psi^0 \rangle \neq 0$. However, this is impossible because $\xi^T \perp \psi^0$, which substantiates our claim.

In the energy representation, the inequality (21) can be written as follows:

$$\sum_{mk} \sum_{nl} \xi_{mk}^{T*} \exp\left(\frac{i}{\hbar} E_m t\right) B_{mk,nl}^j \exp\left(-\frac{i}{\hbar} E_n t\right) \psi_{nl}^0 \neq 0. \tag{22}$$

The condition of local controllability defined by the inequalities (21) and (22) has a clear physical sense. In order to transfer the system from a state to another state in the vicinity of the former state (by means of control), it is necessary and sufficient that there be an instant when the corresponding matrix element of at least one perturbing operator be non-zero.

Instead of investigating the conditions for complete controllability in a small vicinity of the initial state ψ^0, it is more convenient to explain what the following opposite condition means: there is a non-zero vector $\xi^T \perp \psi^0$ which is impossible to arrive at by means of any control. In compliance with the necessary and sufficient condition of local controllability obtained above, the existence of such a vector ξ^T for any $j \in [1, n]$ and $t \in [0, T]$ requires that

$$\sum_{mk} \sum_{nl} \xi_{mk}^{T*} \exp\left(\frac{i}{\hbar} E_m t\right) B_{mk,nl}^j \exp\left(-\frac{i}{\hbar} E_n t\right) \psi_{nl}^0 \equiv 0. \tag{23}$$

Since, by construction, there are no identical energy levels (the degeneracy was accounted for by the introduction of additional quantum numbers), all the functions $\exp(iE_m t/\hbar)$, $m = 0, 1, \ldots$ are linearly independent. Therefore the identity (23) splits up into a number of equations. In the absence of complete controllability, these equations must be satisfied simultaneously for every j, m and n by sets of numbers ξ_{mk}^{T*} constituting a non-zero vector ξ^{T*}. Combined with the orthogonality condition, this gives the homogeneous system of equations

$$\sum_m \sum_k \psi_{mk}^0 \xi_{mk}^{T*} = 0, \tag{24}$$

$$\sum_k \sum_l B_{mk,nl}^j \psi_{nl}^0 \xi_{mk}^{T*} = 0 \quad (j = 1, \ldots, n_1; \; m, n = 0, 1, \ldots) \tag{25}$$

in ξ_{mk}^{T*}, which system can be found from (23) if each coefficient of the linearly independent functions $\exp\left(\frac{i}{\hbar} E_m t\right), \exp\left(-\frac{i}{\hbar} E_n t\right)$ is set equal to zero. As to the controllability in the vicinity of ψ^0, it is necessary and sufficient that this system have only the trivial solution. Let us remark that if the matrices cannot be reduced to block diagonal form, then the verification of this condition can prove to be cumbersome.

We now turn to equations (1.13). Assume that the matrices of the operators $\widehat{B}^j(t)$ in the energy representation have a block diagonal form and that only a finite number of blocks do not have a purely diagonal form; then the problem of local controllability is reduced to the evaluation of the solvability of the finite-dimensional system (24), (25) by finding its rank, since the commutators of purely diagonal submatrices are zero. The problem of whether or not the matrices of the perturbing operators can be reduced to block diagonal form is closely related to the dimensions and the number of irreducible representations of the symmetry group of the unperturbed system and perturbations (Landau and Lifshits, 1963).

In the particular case of non-degenerate energy levels, of practical importance, equations (24), (25) become much simpler because the sums over k and l disappear: $\sum_m \psi_m^0 \xi_m^{T*} = 0$, $B_{mn}^j \psi_n^0 \xi_m^{T*} = 0$, $m, n = 0, 1, 2, \ldots$. Suppose that these equalities hold for any pair of mutually orthogonal vectors, including the vectors of stationary states. Then all the $\widehat{B}_{mn}^j$ with $m \neq n$ must be identical, that is the perturbing matrices are diagonal, and the system is uncontrollable. Conversely, if for some j $\widehat{B}_{mn}^j \neq 0$ for any m, n, $m \neq n$, the solutions of the above system can only be trivial and the quantum system is locally completely controllable.

In the more general case when $B_{mk,nl}^j = b_{mk}^j \delta_{mn} \delta_{kl}$ is a purely diagonal matrix, the stationary states are completely controllable because equations (24), (25) reduce to the trivial equality $\xi_{m_0 k_0}^T = 0$ for one of their components, while the other components of the vector ξ^T can be non-zero. The same conclusion was arrived at above (it followed from the commutativity of the operators $\widehat{B}^j$ and $\widehat{H}^0$.

Generally, no conclusion on the complete controllability of a non-linear system can be made from local controllability or uncontrollability. However, if one can show that the radius of the vicinity of LT-controllability (where L is control intensity and T is time) in the proximity of each of the possible states is no less than a certain number r that is independent of the choice of the initial state, then the system is completely controllable. In fact, any point on the unit sphere will be accessible from any initial state after a finite number of steps $s < \pi/r$ and the whole sphere can be covered by a finite r-net of neighbourhoods of local controllability. In particular, this plays an important role in controllability problems in the Born approximation.

2.3. Global Asymptotic Controllability of Pure States

A system is said to possess *global asymptotic controllability* if for any pair of the initial and final states ψ^0 and ψ^T there exists a sequence of controls $u_n'(t)$, whose values belong to a closed domain Ω such that the sequence of vectors ψ_n^T of final states can be reduced in norm to the vector ψ^T. The physical meaning of this is that the probability that the desired state is attainable approaches unity. The transition time is not limited and can be arbitrarily large but finite.

Using the technique substantiated from the viewpoint of the physics in Landau and Lifshits (1963), we can show that this possibility occurs under general enough assunptions. We shall consider only simple energy equations and a single control function of the form

$$u(t) = \epsilon(a \exp(-i\omega_{12}t) + a^* \exp(-i\omega_{21}t)), \tag{1}$$

where $\omega_{12} = -\omega_{21} = \frac{1}{\hbar}(E_1 - E_2)$ are the Bohr frequencies of transitions from a state 1 to a state 2 and back, ϵa is the amplitude of the control and $0 < \epsilon \le 1$ is a small parameter. If $|a| < 1$, the control function's range of values is evidently bounded.

The system (1.20) in the energy representation with one control function and a simple energy spectrum acquires the form

$$\frac{d\psi_m}{dt} = \frac{i}{\hbar}u(t)\sum_n \exp\left[\frac{i}{\hbar}(E_m - E_n)t\right]B_{mn}\psi_n.$$

We write down just two equations corresponding to the states numbered 1 and 2. This involves no loss of generality if the numbering does not imply an increase in energy levels. On the right hand side, we present only the terms associated with the chosen states while the others will be taken into account as rapidly oscillating remainder terms $i\epsilon o_1(\epsilon, t)$, $i\epsilon o_2(\epsilon, t)$. Thus we have

$$\begin{aligned}
\dot{\psi}_1 &= i\epsilon\nu\psi_2 + i\epsilon o_1(\epsilon, t), \\
\dot{\psi}_2 &= i\epsilon\nu^*\psi_1 + i\epsilon o_2(\epsilon, t),
\end{aligned} \tag{2}$$

where $\nu = (1/\hbar)aB_{12}$. Assume that the system at the initial instant $t = 0$ is in the state 1 which is stationary in the absence of control since it corresponds to a definite value of energy E_1. At this instant $\psi_1(0) = 1$ (the phase is inessential) and $\psi_2(0) = 0$. On the strength of equations (2), $\dot{\psi}_1(0) = 0$, $\dot{\psi}_2(0) = i\epsilon\nu^*$, because $o_2(\epsilon, 0) = 0$. Incidentally, we can also write $o_1(\epsilon, 0) = \frac{\partial o_2}{\partial t}(\epsilon, 0) = 0$.

Let us integrate the system (2) in a formal way at the indicated initial conditions. We shall first eliminate ψ_1, which gives the following equation in ψ_2:

$$\ddot{\psi}_2 + \epsilon^2|\nu^2|\psi_2 = i\epsilon\frac{\partial o_2(\epsilon, t)}{\partial t} - \epsilon^2\nu^* o_1(\epsilon, t). \tag{3}$$

In view of the remarks above, the right hand side equals zero at $t = 0$. Therefore the particular solution of equation (3) satisfying the given initial conditions can be presented as

$$\psi_2 = \frac{i\nu^*}{|\nu|}\sin\epsilon|\nu|t + \frac{1}{|\nu|}\int_0^t\left[i\frac{\partial o_2(\epsilon, \tau)}{\partial\tau} - \epsilon\nu^* o_1(\epsilon, \tau)\right]\sin\epsilon|\nu|(t - \tau)d\tau.$$

Integrating by parts the first term of the integrand on the right hand side, we have:

$$\int_0^t \frac{\partial o_2(\epsilon, \tau)}{\partial \tau} \sin \epsilon |\nu| (t - \tau) d\tau = [o_2(\epsilon, \tau) \sin \epsilon |\nu| (t - \tau)]_{\tau=0}^{\tau=t} +$$

$$+\epsilon |\nu| \int_0^t o_2(\epsilon, \tau) \cos \epsilon |\nu| (t - \tau) d\tau = \epsilon |\nu| \int_0^t o_2(\epsilon, \tau) \cos \epsilon |\nu| (t - \tau) d\tau,$$

and we can reduce the solution to the form

$$\psi_2 = i \frac{\nu^*}{\nu} \sin \epsilon |\nu| t - \epsilon \frac{\nu^*}{\nu} \int_0^t o_1(\epsilon, \tau) \sin \epsilon |\nu| (t - \tau) d\tau +$$

$$+ i\epsilon \int_0^t o_2(\epsilon, \tau) \cos \epsilon |\nu| (t - \tau) d\tau.$$

Since the remainder terms $o_1(\epsilon, t)$ and $o_2(\epsilon, t)$ rapidly oscillate with time (if none of the Bohr frequencies are multiples of ω_{12}), therefore the corresponding integrals converge and are homogeneously bounded in ϵ under suitably weak assumptions. As $\epsilon \to 0$, we can neglect the last two terms on the right hand side. Thus we obtain

$$\psi_2 = i \frac{\nu^*}{\nu} \sin \epsilon |\nu| t + \epsilon o_3(\epsilon, t),$$

where $o_3(\epsilon, t)$ is a homogeneously bounded function of time. The probability w_2 of arriving at the state 2 can be estimated as

$$w_2 = |\psi_2|^2 = \sin^2 \epsilon |\nu| t + \epsilon o(\epsilon, t),$$

where $o(\epsilon, t)$ is also a homogeneously bounded variable. This result shows that if the equalities $t = \pi/2\epsilon|\nu|$ hold, then the probability of reaching the final state tends to unity as $\epsilon \to 0$.

We emphasize once again that this conclusion is valid only in the absence of other transitions with frequencies that are multiples of the selected transition frequency and therefore of the control frequency selected to achieve sharp resonance; otherwise this result is generally invalid.

Now we shall demand that there should be the possibility of controlled transition for one step, that is, without variation of control frequency, from any initial to any final state. Apart from requiring that all the non-diagonal matrix elements of the perturbing operator be non-zero, we therefore demand in addition that there must be no multiples among the Bohr transition frequencies. The energy levels must be incommensurable, or in other words, they must be expressed by coprime numbers. The corresponding control must be a combination of expressions of the form (1) for all possible transition frequencies in the given system.

This conclusion is quite similar to the result concerning the controllability of the classical set of linear oscillators by means of one control function (Begimov et al., 1982).

Concluding this discussion, we note that the transition from one state to another can also be achieved in several steps, that is, it can be accompanied by changes in the spectrum of the control function frequencies. It is sufficient for the multi-step transition that the graph of transitions controlled in one step be connected, that is, such that any two stationary states of the unperturbed system can be connected by a finite number K of arcs, K being common for the whole system.

2.4. Control of the Electron in a Rectangular Potential Well

Suppose that an electron is in a rectangular potential well. The width of the well is a, the walls are infinitely high, and the bottom is inclined. The Hamiltonian of the electron is then

$$\widehat{H} = \widehat{H}^0 + \widehat{ex}E(t),$$

where $\widehat{H}^0 = (\hat{p}^2/2m_e) + U(x)$ is the unperturbed energy operator expressed in terms of the momentum operator $\hat{p} = (\hbar/i)(\partial/\partial x)$; e and m_e are the charge and mass of the electron and $E(t)$ is the strength of the controlling electric field. The role of the operator $\widehat{B}^j$ in (1.1) is played by the coordinate $\hat{x}$, and the control function is the force acting on the electron: $u(t) = -eE(t)$.

The unperturbed potential function of the well can be presented in the form

$$U(x) = \begin{cases} \infty & \text{for } x > a \text{ or } x < 0 \\ 0 & \text{for } 0 < x < a \end{cases}$$

Stationary solutions of the unperturbed problem are defined by the spectrum of the eigenvalues and eigenfunctions of the operator $\widehat{H}^0$, which are known (Landau and Lifshits, 1963) to have the form

$$E_n = \frac{\pi^2\hbar^2}{2m_e a^2}n^2, \quad \psi_n = \sqrt{\frac{2}{a}}\sin\frac{\pi n}{a}x, \quad n = 1, 2, \dots$$

Therefore it makes sense to talk about control of definite energy values.

The eigenfunctions of the momentum operator are found as the solutions of the equation

$$\frac{\hbar}{i} \frac{\partial \phi(x)}{\partial x} = p\phi(x)$$

which are normalized to unity. The boundary conditions at the well walls are zero. It follows that the momentum has the same eigenfunctions as the energy (the operators of these quantities commute). Notice that for any state with a given energy E there are two equally probable values of momentum: $p_n^{\pm} = \pm\sqrt{2mE_n}$.

Let us consider a state in which the electron is almost certainly at a point with the coordinate x' within the well. Naturally, this state is idealized because it requires an infinite spectrum of energies for it. However, this does not contradict the assumed geometry of the well with infinitely high walls. We can employ the following identity for the energy eigenfunctions introduced above:

$$\sum_n \psi_n^*(x')\psi_n(x) = \delta(x - x');$$

this identity can be easily verified. Its right hand side can be treated as the unnormalized amplitude of probability that the electron is at the point x' under the condition that it is in a state with the certain coordinate x' indicated above. Then the left hand side of the identity means that the Fourier coefficients of the state in question in the energy basis have a form similar to the basis functions themselves, that is,

$$\psi_n(x') = \sqrt{\frac{2}{a}} \sin\frac{\pi n}{a}x'.$$

To bring the electron to a new state with a certain coordinate x'' means to arrive at a state with the expansion coefficients

$$\psi_n(x'') = \sqrt{\frac{2}{a}} \sin\frac{\pi n}{a}x''.$$

Consequently, the formulation of the problem of taking the electron from one point to another inside the well is quite correct within the framework of our idealization.

The matrix elements of the coordinate operator in the energy representation can be found from the formulas

$$x_{nn'} = \langle n|\hat{x}|n'\rangle = \int_0^a \psi_n^*(x)x\psi_{n'}(x)dx = \frac{2}{a}\int_0^a x\sin\frac{\pi n x}{a}\sin\frac{\pi n' x}{a}dx$$

and therefore are

$$
x_{nn'} = \begin{cases} \dfrac{2a}{\pi^2}\left[\dfrac{1}{(n-n')^2} - \dfrac{1}{(n+n')^2}\right] & (n-n' \neq 0 \text{ is odd}), \\ 0 & (n-n' \neq 0 \text{ is even}), \\ a/2 & (n-n' = 0). \end{cases}
$$

To investigate global asymptotic controllability, we plot the graph of possible one-step transitions whose vertices correspond to states with a definite energy and whose arcs correspond to non-zero elements of the matrix (Figure 2.4.1). It is clear from the graph that not every state is accessible from any initial state in one step. However, this is a connected graph and any state is accessible from any initial state in two steps at the most. Therefore, the controllability depends on relations between the energy levels of the system. In this case they are related as squares of integers. That is why there are final states that are not asymptotically accessible from any initial state in a finite number of steps. The transition is only possible between states that are superpositions of states with coprime squares of quantum numbers.

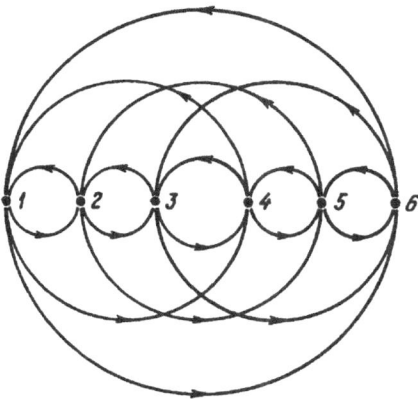

Fig. 2.4.1

However, the wave functions $\psi_n(x)$ of the given form with every possible n such that the numbers n^2 are coprime do not comprise a complete system of functions forming a basis in $L_2[0, a]$. It follows that in the case of a rectangular potential well it is impossible to control the coordinate x of the electron and maintain its arbitrarily precise localization. (A correction must be made therefore in Butkovskiy and Samoilenko, 1979b). But if the potential is supplemented by a perturbation, even a small one, which would render every energy level rationally independent of each other (Flügge, 1971), the completeness of the system of attainable stationary states is restored, and arbitrary asymptotically precise localization of the electron becomes possible.

2.5. Control of a Two-Spin System

Assume that there are two electrons in a magnetic field. The electrons are on the z-axis at a distance $\pm a$ from the origin. The magnetic field consists of (1) a constant component equal to B_z^0 at the origin and possessing a gradient $\bar{e}_z B_z'$ that is constant within the closed interval $[-a, a]$, and (2) a controlled homogeneous transverse component $\bar{e}_x B_x(t)$. Here $\bar{e}_x$ and $\bar{e}_z$ are the basis vectors in the Cartesian system of coordinates. The shift of unperturbed energy levels due to the interaction between the electrons will be considered inessential in this case, which is quite acceptable for sufficiently large a.

The Hamiltonian of the system is the sum of the Hamiltonians of the two electrons: $\widehat{H} = \widehat{H}_1 + \widehat{H}_2$, where

$$\widehat{H}_1 = -\mu[\hat{\sigma}_z^1 B_z^1 + \hat{\sigma}_x^1 B_x(t)],$$
$$\widehat{H}_2 = -\mu[\hat{\sigma}_z^2 B_z^2 + \hat{\sigma}_x^2 B_x(t)],$$
$$B_z^1 = B_z^0 - aB_z', \quad B_z^2 = B_z^0 + aB_z'$$

and

$$\hat{\sigma}_x^1 = \hat{\sigma}_x^2 = \begin{pmatrix} 0 & 1 \\ 1 & 0 \end{pmatrix}, \quad \hat{\sigma}_z^1 = \hat{\sigma}_z^2 = \begin{pmatrix} 1 & 0 \\ 0 & -1 \end{pmatrix}$$

are the operators of spins 1 and 2 of the particles, and μ is the magnetic moment of the electron. The basis states will be assumed to be such that they correspond to certain energy values in the unperturbed state, when each of the spins is either parallel or antiparallel to the unperturbed field. To number the components of the state vector we have to use two superscripts because the complete set of physical quantities defining the system includes the spins of both particles. These can be measured independently because the electrons do not interact by definition. Each of the spin operators has a superscript and can only change the state vector related to the corresponding particle.

The subscripts of the state vector-function ψ_{mn} can assume the values 0 and 1 respectively for the lesser and greater energy of the interaction between the electron magnetic moment and the field. The subscript m corresponds to the electron at the point $z = -a$ and the subscript n corresponds to the one at $z = a$.

Let us compile the table of products between the operators and the vectors:

$$\hat{\sigma}_x^1 \psi_{0n} = \psi_{1n}, \quad \hat{\sigma}_x^1 \psi_{1n} = \psi_{0n},$$

$$\hat{\sigma}_z^1 \psi_{0n} = \psi_{0n}, \quad \hat{\sigma}_z^1 \psi_{1n} = -\psi_{1n},$$

$$\hat{\sigma}_x^2 \psi_{m0} = \psi_{m1}, \quad \hat{\sigma}_x^2 \psi_{m1} = \psi_{m0},$$

$$\hat{\sigma}_z^2 \psi_{m0} = \psi_{m0}, \quad \hat{\sigma}_z^2 \psi_{m1} = -\psi_{m1},$$

$$m, n = 0, 1.$$

In compliance with this table, we can construct the Hamiltonian matrices for the two electrons:

$$\hat{H}_1 = -\mu(B_z^0 - aB_z') \begin{pmatrix} 1 & 0 \\ 0 & -1 \end{pmatrix} - \mu B_x(t) \begin{pmatrix} 0 & 1 \\ 1 & 0 \end{pmatrix},$$

$$\hat{H}_2 = -\mu(B_z^0 + aB_z') \begin{pmatrix} 1 & 0 \\ 0 & -1 \end{pmatrix} - \mu B_x(t) \begin{pmatrix} 0 & 1 \\ 1 & 0 \end{pmatrix},$$

The unperturbed operator in this form is diagonal, the control for the two electrons is $u(t) = \mu B_x(t)$, and the operators that are energy conjugates of control are presented as the spin matrix σ_x with non-zero non-diagonal elements.

The system has the following energy levels:

$$E_1 = -2\mu B_z^0, \quad E_2 = -2\mu a B_z', \quad E_3 = 2\mu a B_z', \quad E_4 = 2\mu B_z^0.$$

Transitions between these levels can be carried out by controlling each subsystem with the same $u(t)$. The Bohr transition frequencies for the two electrons are different:

$$\omega_1 = \frac{\mu}{\hbar}(B_z^0 - aB_z'), \quad \omega_2 = \frac{\mu}{\hbar}(B_z^0 + aB_z').$$

It is obvious that the conditions of global asymptotic controllability are satisfied. By exciting the system by the first transition frequency, we can transfer the first electron into any of its two stationary states. The same can be said of the second electron. Using the two frequencies in various combinations, we can perform the asymptotic transition from any state into any other in one step of sufficient duration. Reduction of the transition time is the problem of optimization, the methods of which are discussed in Chapter 4.

Applying staircase control, we can integrate the system accurately and then find switch moments by matching the solutions and the boundary conditions. However, this entails rather cumbersome calculations that are inappropriate here because the algebraic methods discussed in Chapter 3 are more efficient.

2.6. Finite Control of a Particle Spin State

Two-level systems appear to be among the simplest quantum-mechanical objects for which the problems of controlling pure states can be considered. Many important physical systems are considered within the framework of the theory of this type of system in Dirac (1958) and Landau and Lifshits (1963). Let us take spin to be the characteristic of the state of a controlled system. Assume that ψ_1 and ψ_2 are its basis states in which the projections of spins on the z-axis are $\hbar/2$ and $-\hbar/2$ respectively. Then the state ψ as a function of time t can be described by the superposition

$$\psi = \psi(t) = a_1(t)\psi_1 + a_2(t)\psi_2, \quad t \geq 0, \tag{1}$$

where $a_1(t)$ and $a_2(t)$ are probability amplitudes of the states ψ_1 and ψ_2 respectively. The condition of normalization has the following form:

$$a_1^*(t)a_1(t) + a_2^*(t)a_2(t) = |a_1(t)|^2 + |a_2(t)|^2 = 1, \quad t \geq 0. \tag{2}$$

The motion of the system in a magnetic field is described by the equations for the amplitudes (Blokhintsev, 1981b)

$$\dot{a}_1 = i\omega(a_1 \cos\theta + a_2 e^{-i\phi}\sin\theta),$$
$$\dot{a}_2 = i\omega(a_1 e^{i\phi}\sin\theta - a_2 \cos\theta), \tag{3}$$

where $\omega = \mu B/\hbar$, μ is the magnetic moment of the microparticle, B is the absolute value of the magnetic field strength vector, and θ and ϕ are the spherical coordinates of the radius-vector B.

The vector B can be regarded as the control. Therefore the parameters ω, ϕ, θ can be viewed as controls; they are generally functions of time t. There are natural constraints for these parameters:

$$\omega = \omega(t) \geq 0, \ 0 \leq \theta \leq \pi, \ -\pi < \phi \leq \pi. \tag{4}$$

The normalization conditions (2) for equations (3) hold automatically by virtue of the general laws of motion which, incidentally are easily verified by calculating $\frac{d}{dt}[a^*(t), a(t)]$ and taking into account equations (3), their conjugate equations and the initial conditions

$$a_1(0) = \alpha_1, \ a_2(0) = \alpha_2, \ |\alpha_1|^2 + |\alpha_2|^2 = 1, \ a = (a_1, a_2). \tag{5}$$

Various problems can be considered for this system: those of optimization, controllability, finite control, etc. Let us consider the following problem of finite control (Appendix 2). Suppose that equations (3) have the initial conditions (5). Find a time function $\omega = \omega(t)$ and constant parameters θ and ϕ such that at the instant $t = T$ the following final conditions hold:

$$a_1(T) = \beta_1, \ a_2(T) = \beta_2, \tag{6}$$

where β_1 and β_2 are preset numbers that satisfy the conditions of normalization

$$|\beta_1|^2 + |\beta_2|^2 = 1. \tag{7}$$

In other words, find both a time-independent direction of the magnetic field in space and a time-dependent amplitude of this field such that the system is transferred from the initial state to the desired final state during a certain interval.

We now turn to the solution of this. First of all, we can replace the independent variable t by τ using

$$\tau = \int_0^t \omega(s)ds, \tag{8}$$

since in view of (4), τ is always positive. Then (3) can be rearranged as

$$\dot{a}_1(\tau) = ia_1 \cos\theta + ia_2 e^{-i\phi} \sin\theta,$$
$$\dot{a}_2(\tau) = ia_1 e^{i\phi} \sin\theta - ia_2 \cos\theta. \tag{9}$$

The eigenvalues of this system are $\lambda_1 = i$ and $\lambda_2 = -i$. The initial values of $a_1(\tau)$ and $a_2(\tau)$ remain the same (5): $\tau = 0$ corresponds to $t = 0$. However, the value $\tau = \tau_1$ that corresponds to the final instant $t = T$ can no longer be preset, and therefore τ_1 can also be regarded as a control parameter whose value must be determined from the solution of the problem. The parameter

τ_1 appears instead of the required function $w(t)$. Any function $w(t)$ can be regarded as the one sought if it satisfies the condition

$$w(t) \geq 0, \quad \int_0^T w(t)dt = \tau_1. \tag{10}$$

Therefore the problem has been reduced to that of finding three numbers θ, ϕ, and τ_1 for which the conditions

$$a_1(\tau_1) = \beta_1, \quad a_2(\tau_1) = \beta_2 \tag{11}$$

are satisfied in view of (9).

To solve this problem we integrate (9) given the initial conditions (5) and using well-known methods. In matrix notation, the solution has the form

$$a(\tau) = S(\tau, \theta, \phi)\alpha, \tag{12}$$

where

$$a(\tau) = \begin{pmatrix} a_1 & (\tau), \\ a_2 & (\tau) \end{pmatrix}, \quad \alpha = \begin{pmatrix} \alpha_1 \\ \alpha_2 \end{pmatrix}, \quad S(\tau, \theta, \phi) = \begin{pmatrix} u & -v^* \\ v & u^* \end{pmatrix}$$

and

$$u = \cos\tau + i\sin\tau\cos\theta, \quad v = i\sin\tau\sin\theta e^{i\phi}. \tag{13}$$

It follows from the general theory that this transformation of α into $a(t)$ is unitary. It is easy to verify this. The determinant of the matrix S in (12) equals unity.

In this example we have managed to find an explicit form for the unitary transformation S defining the motion of the quantum system and the dependence of S on the control parameters τ, θ, and ϕ. Setting $\tau = \tau_1$ in (12) and using the final conditions (11), we arrive at the system of algebraic equations

$$\alpha_1 u_1 - \alpha_2 v_1^* = \beta_1, \quad \alpha_2 u_1^* + \alpha_1 v_1 = \beta_2, \tag{14}$$

where $u_1 = u(\tau_1)$ and $v_1 = v(\tau_1)$. Replacing all the variables in the second equation (14) by their complex conjugates, we arrive at the system of equations

$$\alpha_1 u_1 - \alpha_2 v_1^* = \beta_1, \quad \alpha_2^* u_1 + \alpha_1^* v_1^* = \beta_2^*. \tag{15}$$

This is a system of linear algebraic equations in $u_1 - v_1^*$. It is always solvable because its determinant equals -1 by virtue of the normalization conditions. Hence

$$u_1 = \beta_1 \alpha_1^* + \beta_2^* \alpha_2, \quad v_1 = \beta_2 \alpha_2^* - \beta_1^* \alpha_2, \quad -v_1^* = \beta_1 \alpha_2^* - \beta_1^* \alpha_1. \qquad (16)$$

Substituting these values of u_1 and v_1 into (13) at $\tau = \tau_1$, we arrive at the following system of transcendental equations in τ_1, θ and ϕ needed to solve this problem of finite control:

$$\cos \tau_1 + i \sin \tau_1 \cos \theta = u_1, \qquad (17)$$

$$i \sin \tau_1 \sin \theta e^{-i\phi} = v_1. \qquad (18)$$

Separating the real and imaginary parts in (17), we immediately find that

$$\sin \tau_1 = \pm\sqrt{1 - c^2}, \quad \cos \tau_1 = c,$$

$$\sin \theta = \sqrt{\frac{1 - A^2}{1 - c^2}} \geq 0, \quad \cos \theta = \frac{\pm D}{\sqrt{1 - c^2}}, \qquad (19)$$

where

$$c = \operatorname{Re} u_1, \quad D = \operatorname{Im} u_1, \quad A = |u_1| \qquad (20)$$

and u_1 is determined in (16). The appropriate choice of the combination of signs in the expressions for $\sin \tau_1$ and $\cos \theta$ is found from the equality

$$\operatorname{sgn}(\cos \theta \sin \tau_1) = \operatorname{sgn} D. \qquad (21)$$

Thus for $D \geq 0$, the solutions for τ_1 and θ satisfy the equations

$$\sin \tau_1 = \sqrt{1 - c^2} \geq 0, \quad \sin \tau_1 = -\sqrt{1 - c^2} \leq 0,$$

$$\cos \theta = \frac{D}{\sqrt{1 - c^2}} \geq 0, \quad \cos \theta = \frac{-D}{\sqrt{1 - c^2}} \geq 0. \qquad (22)$$

For $D \leq 0$, the solutions satisfy the equations

$$\sin \tau_1 = \sqrt{1 - c^2} \geq 0, \quad \sin \tau_1 = -\sqrt{1 - c^2} \leq 0,$$

$$\cos \theta = \frac{-D}{\sqrt{1 - c^2}} \leq 0, \quad \cos \theta = \frac{D}{\sqrt{1 - c^2}} \geq 0. \qquad (23)$$

Comparing the arguments of the numbers in (18), we can easily find the parameter ϕ:

$$\phi = \begin{cases} \frac{\pi}{2} - \arg v_1, & \text{if } \sin \tau_1 > 0, \\ -\frac{\pi}{2} - \arg v_1, & \text{if } \sin \tau_1 < 0, \end{cases} \tag{24}$$

where v_1 is defined in (16). Note that the normalization conditions (7) and (2) ensure that the system (17) is consistent and the values of its solutions are real.

Thus equations (19), (24) give the complete solution of the finite control problem posed above. Here the variation of the amplitude of the controlling magnetic field can be any function of time $\omega(t) \geq 0$ provided that

$$\int_0^T \omega(t)dt = \tau_1. \tag{25}$$

Suppose that the initial state, for instance, is such that $\alpha_1 = 1$ and $\alpha_2 = 0$. The strength of the controlling magnetic field is bounded by $B_{\max}$. It is required to find the direction of the field and the minimum time $T_{\min}$ such that the system is transferred into a new state $\beta_1 = 0$, $\beta_2 = 1$. Let us solve the problem.

We can find from (20) that $C = 0$ and $D = 0$. On the strength of (19), $\cos \tau_1 = 0$. Hence $\tau_{1\min} = \pi/2$. Therefore in view of (25) we have

$$\int_0^{T_{\min}} \frac{\mu B_{\max}}{\hbar} dt = \frac{\pi}{2}. \tag{26}$$

Hence

$$T_{\min} = \frac{\pi \hbar}{2\mu B_{\max}}. \tag{27}$$

We now find from (19) that $\cos \beta = 0$, therefore $\theta = \pi/2$. Finally, (24) gives

$$\phi = \pi/2. \tag{28}$$

Thus the system will be transferred into the desired state if the maximum field $B_{\max}$ is directed oppositely with respect to the y-axis and acts during time $T_{\min}$ defined by (27). Note that if there are short-term yet strong impulses of the magnetic field, that is,

$$B(t) = B_0 \delta(t), \tag{29}$$

where B_0 is the 'area' of the impulse $B(t)$, then the time T of the transition process (the interval of finite control) can be made arbitrarily small by making B_0 large enough. Substituting (29) into (25), we can obtain

$$\int_0^T \frac{\mu B_0}{\hbar} \delta(t) dt = \frac{\pi}{2}. \tag{30}$$

Hence the required 'area' B_0 of the impulse B directed along the negative y-axis is

$$B_0 = \frac{\pi \hbar}{2\mu}. \tag{31}$$

Taking the same example, let us find the control, that is, the angles θ and ϕ, if the problem of finite control of the electron spin is not posed in so much detail; namely, we are not interested this time in the amplitudes and therefore the phases of the states, but restrict ourselves to setting the probabilities that the final states will be arrived at. Therefore we demand that the transition from the initial state $\alpha_1 = 1$, $\alpha_2 = 0$ to the state (β_1, β_2) must be carried out merely under the condition that $|\beta_2| = 1$ (hence $\beta_1 = 0$).

Separating the real and imaginary parts in (17), it then follows immediately that

$$\cos \tau = 0, \tag{32}$$

$$\sin \tau \cos \theta = 0. \tag{33}$$

Taking the modulus in equation (18), we also see immediately that the condition

$$|\sin \tau \sin \theta| = 1 \tag{34}$$

must hold. This last equality clearly ensures that (33) will hold. Thus the solution of the problem under consideration is that the vector of the controlling magnetic field must be orthogonal to the z-axis.

It is interesting that, unlike the case where it was required to obtain the amplitudes of the state ($\beta_1 = 0$, $\beta_2 = 1$), in this case, where it is only required to obtain the probability that the state ($\beta_1 = 0$, $\beta_2 = 1$) is attained, the angle ϕ is arbitrary, that is, the required control, as expected, does not depend on the angle.

2.7. Control of Quantum Averages of Physical Quantities

The general problem of control of the averages of physical quantities (observables) was formulated above, and approaches to its solution were outlined. We shall consider below the case where the equations of motion for the observables coincide exactly with the respective classical equations.

Suppose that the Hamiltonian H has a quadratic-linear form whose coefficients depend on $u = u(t)$:

$$H(q,p,t,u) = \sum_{\substack{k=1 \\ m=1}}^{n} [a_{km}(t,u)q_k q_m + b_{km}(t,u)q_k p_m + c_{km}(t,u)p_k p_m] +$$

$$+ \sum_{m=1}^{n} [d_m(t,u)q_m + e_m(t,u)p_m] + g(t,u). \qquad (1)$$

In matrix notation, this Hamiltonian can be rewritten as

$$H(q,p,t,u) = H(l,t,u) = (l', F(t,u)l) + (l', f(t,u)) + g, \qquad (2)$$

where $l = (q,p)$, $F(t,u)$ is a quadratic matrix of dimension $2n \times 2n$ and $f(t,u)$ is a column matrix of dimension 2. Then, in compliance with the Ehrenfest theorem (Blokhintsev, 1981b) and Appendix 1, the equations of motion for the averages $\bar{q}_k(t)$ and $\bar{p}_k(t)$ coincide *exactly* with the classical Hamiltonian equations which are linear in q_k and p_k in this case:

$$\dot{\bar{q}}_k = \frac{\partial H(\bar{q},\bar{p},t,u)}{\partial \bar{p}_k}, \quad \dot{\bar{p}}_k = -\frac{\partial H(\bar{q},\bar{p},t,u)}{\partial \bar{q}_k}, \qquad (3)$$

or

$$\dot{\bar{q}}_k = \sum_{m=1}^{n} [a_{km}(t,u) + a_{mk}(t,u)]\bar{q}_m + \sum_{m=1}^{n} b_{km}(t,u)\bar{p}_m + d_k(t,u),$$

$$\dot{\bar{p}}_k = -\sum_{m=1}^{n} [c_{km}(t,u) + c_{mk}(t,u)]\bar{p}_m - \sum_{m=1}^{n} b_{mk}(t,u)\bar{q}_m + e_k(t,u). \qquad (4)$$

Therefore when the Hamiltonian is a quadratic-linear form with respect to the canonically conjugate variables $\bar{q}_k$ and $\bar{p}_k$, the equations of motion in the quantum averages of these variables are ordinary linear differential equations with generally variable coefficients depending on time and controls. It is characteristic that functions depending directly on control can enter these equations both additively and multiplicatively.

It is of importance that control problems for this kind of system can be solved by means of the classical theory of control of lumped parameter systems. However, there are usually great difficulties in the optimization of the systems with controls in the coefficients.

A quantum oscillator is an illustrative example of a system with a quadratic Hamiltonian. The investigation of control of a quantum harmonic oscillator is very useful primarily because the oscillator Hamiltonian occurs in many problems of quantum physics and in particular where oscillations occur: in the theory of molecular and crystal oscillations, quantum field theory, etc. (Akhiezer and Berestetskii, 1969; Bogolubov and Shirkov, 1976; Karlov and Malenkov, 1966). For instance, take a charged linear micro-oscillator in superimposed electric and magnetic fields: it can be easily obtained from the system (3) that the equations of motion in the averages of the coordinates $\bar{q}_1, \bar{q}_2, \bar{q}_3$ of the micro-oscillator position are

$$\ddot{\bar{q}}_1 = -\left(\frac{k}{m}\right)\bar{q}_1 + \left(\frac{e}{me}\right)[B_3(t)\dot{\bar{q}}_2 - B_2(t)\dot{\bar{q}}_3] + \left(\frac{e}{m}\right)E_1(t) \tag{5}$$

where e is the charge, m is the mass, c is the velocity of light, k is the modulus of elasticity and $E = (E_1, E_2, E_3)$ and $B = (B_1, B_2, B_3)$ are the strengths of the electric and magnetic fields respectively, which play the role of controls. The other two equations of motion can be obtained from equation (5) by means of a cyclic rearrangement of the subscripts 1,2,3.

In particular, we can deduce from (5) the equations for an oscillator in an electric field:

$$\ddot{\bar{q}} + \left(\frac{k}{m}\right)\bar{q} = \left(\frac{e}{m}\right)E(t). \tag{6}$$

This is a linear equation where the control $E(t)$ enters additively. We can pose and solve many control problems for this kind of system using the well-developed apparatus of the theory of optimization, controllability, finite control, etc.

Note that the momentum variance $\sigma_p^2(t)$ does not depend on the control $E(t)$ and is always constant, equal to its initial value $\sigma_p^2(0)$. Indeed, if there is a particle moving in a field with potential $V(q)$, we have

$$\frac{d\overline{(p^2)}}{dt} = \left(\bar{p}\frac{\partial\overline{V}}{\partial q} + \frac{\partial\overline{V}}{\partial q}\bar{p}\right) = -2eE(t)\bar{p}. \tag{7}$$

In view of the equations of motion it follows that

$$\frac{d(\bar{p})^2}{dt} = 2\bar{p}\dot{\bar{p}} = -2\bar{p}\frac{\partial\overline{V}}{\partial q} = -2eE(t)\bar{p}. \tag{8}$$

We can now deduce from the last two equalities that

$$\frac{d\sigma_p^2(t)}{dt} = \frac{d\overline{(\bar{p})^2}}{dt} - \frac{d(\bar{p})^2}{dt} = 0,$$

and hence $\sigma_p^2(t) = \sigma_p^2(0)$, that is, $\sigma_p^2(t)$ is a constant. Using the commutation relations, it is then easy to see that

$$\frac{d^2\sigma_q^2(t)}{dt^2} = \frac{2}{m^2}\sigma_p^2(t) = \frac{2}{m^2}\sigma_{\hat{p}}^2(0), \tag{9}$$

where $\sigma_q^2(t)$ is the variance of the coordinate q of the particle position. Therefore $\sigma_q^2(t)$ does not depend on control either, that is, the variance is completely uncontrollable.

However, one should not make the conclusion that the variance is also uncontrollable in the more general case of an anharmonic oscillator with anharmonic coefficients depending on control. On the contrary, complete controllability is possible here under certain conditions. Then the uncertainty of either the momentum or the coordinate can be made in principle arbitrarily small (but not simultaneously, of course). The consequence of this is of great significance for the design of automata on controlled transitions of quantum systems (see §3.5 below).

2.8. Control of Coherent States of a One-Dimensional Quantum Oscillator by Means of an External Force

The two following sections discuss some approaches to the solution of problems of controlling coherent states of quantum systems. A typical example is a one-dimensional harmonic oscillator and another typical case is a free particle (Butkovskiy and Pustylnikova, 1982). We shall consider both. The appropriate theory can develop, for instance, from the well-known methods of control of classical systems. This section illustrates the point by the problem of controlling a quantum oscillator.

The solution of the problem posed below may serve as an example of reducing the problems of quantum system control to problems of controlling ordinary classical systems with subsequent application of the methods of the classical theory of control. Besides, the case of the oscillator with variable frequency, which is acted upon by an external force, is also an example of a quantum system defined by a partial differential equation, which makes it possible to arrive at an exact solution of the control problem.

The control problem consists in the following. There is a quantum oscillator whose state (wave function) obeys the non-stationary one-dimensional

Schrödinger equation. It is assumed that the strength $u(t)$ of the external field and the eigenfrequency $\omega(t)$ of the oscillator, both entering the Hamiltonian, can be varied. It is required that these functions should be determined such that the quantum system in question will transfer from an initial stationary state into a desired final state during time T with maximum or preassigned probability.

Let us first consider the following problem of controlling a quantum oscillator. Suppose that prior to the instant $t \leq t_0$ the oscillator is in a known stationary state $|n\rangle$. At $t_0 = 0$, which is assumed to be the initial instant, the oscillator is put into a homogeneous field of the strength $u(t)$. It is required to find $u(t)$ such that the quantum oscillator arrives at a desired stationary state $|m\rangle$ at a preset time $T > t_0$ with maximum or preassigned probability.

The Schrödinger equation for an oscillator in an external field has the following form:

$$i\frac{\partial \psi}{\partial t} = -\frac{1}{2}\frac{\partial^2 \psi}{\partial x^2} + \frac{1}{2}\omega_0^2 x^2 \psi - u(t)x\psi \qquad (1)$$

(using the system of units where $m = \hbar = 1$). Without loss of generality we can assume $u(t)$ to be finite on $[0, T]$. Let us find the form of the wave function ψ for the system (1). If we use the following substitution (Baz et al., 1971)

$$\psi(x,t) = e^{i\left(\dot{\eta}x_1 + \sigma(t)\right)}\phi(x_1, t), \quad x_1 = x - \eta(t), \qquad (2)$$

corresponding to the transition to a moving frame, equation (1) can be represented in the form

$$i\frac{\partial \phi}{\partial t} = -\frac{1}{2}\frac{\partial^2 \phi}{\partial x_1^2} + \frac{1}{2}\omega_0^2 x_1^2 \phi + \left(\ddot{\eta} + \omega_0^2 \eta - u(t)\right)x_1 \phi +$$

$$+\left(\dot{\sigma} - \frac{1}{2}\dot{\eta}^2 + \frac{1}{2}\omega_0^2 \eta^2 - u(t)\eta\right)\phi \qquad (3)$$

It follows from (3) that if the relations

$$\ddot{\eta} + \omega_0^2 \eta = u(t), \quad \eta(0) = \dot{\eta}(0) = 0, \qquad (4)$$

$$\sigma(t) = \int_0^t L(t')dt', \qquad (5)$$

where $L(t)$ is the classical Lagrangian

$$L(t) = \frac{1}{2}\dot{\eta}^2 - \frac{1}{2}\omega_0^2 \eta^2 + u(t)\eta, \qquad (6)$$

are valid, then the function $\phi(x,t)$ will satisfy the equation

$$i\frac{\partial\phi}{\partial t} = -\frac{1}{2}\frac{\partial^2\phi}{\partial x_1^2} + \frac{1}{2}\omega_0^2 x_1^2\phi. \tag{7}$$

This is the Schrödinger equation for an oscillator that is free from any external forces. It is easy to see that the solution of equation (1) has the form

$$\psi(x,t) = \phi(x - \eta(t))\exp\left[i\dot{\eta}(x - \eta) + i\int_0^t L(\tau)d\tau\right]. \tag{8}$$

Therefore the evolution of the wave function is characterized by the variables $\eta(t)$ and $L(t)$ relating to the classical oscillator. This makes it possible to consider the problem of control of the classical oscillator corresponding to the problem of control of the quantum oscillator posed above; in this way the solution of the latter is reduced to the solution of the former.

We represent the amplitudes of the forced oscillations $\eta(t)$ in the form

$$\eta = \eta(t) = \frac{1}{\sqrt{2\omega_0}}(e^{i\omega_0 t}d^* + e^{-i\omega_0 t}d). \tag{9}$$

The variable

$$d = d(t) = \frac{i}{\sqrt{2\omega_0}}\int_0^t e^{i\omega_0\tau}u(\tau)d\tau \tag{10}$$

has the following meaning: if we change from the coordinate x and the momentum p to the dimensionless variables $X = \sqrt{\frac{\omega_0}{2}}x$, $P = \frac{1}{\sqrt{2\omega_0}}p$ and set

$$R = X + iP = \frac{\omega_0\eta + i\dot{\eta}}{\sqrt{2\omega_0}}, \tag{11}$$

then there exists a one-to-one correspondence between the points R in the plane and the states of the classical oscillator at each instant t. The variable $d(t)$ is the displacement of this point due to the external force.

Using (9), we find that the state of the classical oscillator at $t = T$ can be described as

$$R_T = \left.\frac{\omega_0\eta + i\dot{\eta}}{\sqrt{2\omega_0}}\right|_{t=T} = de^{-i\omega_0 T}, \tag{12}$$

where

$$d = d(T) \equiv \sqrt{\nu}e^{i\beta}, \tag{13}$$

$\nu = |d|^2$ characterizing the excitation of the quantum oscillator by an external force. Thus the probability of transferring the oscillator from the stationary state $|n\rangle$ to the state $|m\rangle$ has the form (Baz et al., 1971)

$$W_{mn} = \frac{n_1!}{n_2!}\nu^k|L_{n_1}^k(\nu)|^2 e^{-\nu}, \ k = |n - m|, \tag{14}$$

where $n_1 = \min(m, n)$, $n_2 = \max(m, n)$, and $L_n^\alpha(x)$ are the generalized Laguerre polynomials. Equation (14) allows us to define the value $\nu = \nu_0$ such that the probability of transition under the action of the external force $u(t)$ attains its maximum (or a preassigned value).

The state of the corresponding classical oscillator at $t = T$, in view of (12) and (13), can be described as

$$\eta(T) = \frac{\sqrt{2\nu_0}}{\sqrt{\omega_0}}\cos(\omega_0 T - \beta), \tag{15}$$

$$\dot{\eta}(T) = -\sqrt{2\omega_0\nu_0}\sin(\omega_0 T - \beta). \tag{16}$$

Therefore the desired control $u(t)$ for the quantum oscillator must be at the same time a solution of the problem of controlling the classical oscillator: to find a function $u(t)$ transferring the system

$$\ddot{\eta} + \omega_0^2\eta = u(t) \tag{17}$$

from the initial state

$$\eta(0) = \dot{\eta}(0) = 0 \tag{18}$$

to the prescribed final state (15), (16) during time T.

Using the expression of the general solution of the finite control of linear autonomous lumped parameter systems given in its explicit form (Appendix 2), we arrive at the solution of the problem of classical oscillator formulated above:

$$u(t) = \frac{\sqrt{2\nu_0}}{\sqrt{\omega_0}}\delta'(t)\cos\beta + \sqrt{2\nu_0\omega_0}\delta(t)\sin\beta + \omega_0^2\phi(t) + \phi''(t), \tag{19}$$

where $\phi(t)$ is an arbitrary finite function on the interval $[0, T]$. This control (19) is at the same time the solution of the quantum oscillator problem posed above.

By way of illustration, let us consider the following problem. Find a control $u(t)$ that transfers the system (1) from the state $|0\rangle$ to a prescribed state $|m\rangle$ with maximum probability. This probability

$$W_{m0} = \frac{1}{m!} \nu^m e^{-\nu} \tag{20}$$

reaches its maximum when $\nu = m$ and therefore the control solving the problem has the form

$$u(t) = \frac{\sqrt{2m}}{\sqrt{\omega_0}} \cos \beta \delta'(t) + \sqrt{2m\omega_0} \sin \beta \delta(t) + \omega_0^2 \phi(t) - \phi''(t). \tag{21}$$

Now let us consider a more direct approach to the problem of control of a one-dimensional oscillator under the action of an external force employing the Green's functions for (1). It is shown in Feynman and Hibbs (1965) that the Green's function for this case is

$$G(x, \xi, t) = \frac{1}{\sqrt{2\pi i \sin t}} \exp iS, \tag{22}$$

where

$$S = \frac{1}{2 \sin t} \left[(x^2 + \xi^2) \cos t - 2x\xi + 2x \int_0^t u(\tau) \sin \tau d\tau + \right.$$

$$\left. +2\xi \int_0^t u(\tau) \sin(t - \tau) d\tau - 2 \int_0^t \int_0^\tau u(\tau) u(s) \sin(t - \tau) \sin s \, ds \, d\tau \right]. \tag{23}$$

As above, the problem is to find a function (force) $u(t)$ on the interval $[0, T]$ such that the oscillator goes from the initial nth stationary state at $t = 0$ (characterized by the function ψ_n) to the mth stationary state at $t = T$ (characterized by the function $\psi_m(x)$), the probability of the event being maximum.

The function $\psi_k(x)$ of the quantum oscillator has the form

$$\psi_k(x) = (k! 2^k \sqrt{\pi})^{-1/2} H_k(x) \exp\left(-\frac{x^2}{2}\right), \tag{24}$$

$k = 0, 1, 2, \ldots$, where $H_k(x)$ is the kth Hermite polynomial. In the general case, the probability amplitude a_{mn} that such a transition occurs, if the Green's function $G(x, \xi, t)$ is known, has the form

$$a_{mn}(t) = \int_{-\infty}^{\infty} \int_{-\infty}^{\infty} \psi_n^*(x) G(x, \xi, t) \psi_m(\xi) d\xi dx. \tag{25}$$

The explicit form of a_{mn} for the oscillator, taking into account (23) and (24), is cited in Feldbaum (1971) and is given by

$$a_{mn}(t) = \frac{a_{00}(t)}{\sqrt{m!n!}} \sum_{r=0}^{l} \frac{m!}{(m-r)!} \cdot \frac{n!}{(n-r)!} \cdot \frac{1}{r!}[i\beta(t)]^{n-r}[i\beta^*(t)]^{n-r}, \quad (26)$$

where

$$l = \min(m, n), \quad (27)$$

$$a_{00}(t) = \exp\left\{-\frac{1}{2}\int_0^t\int_0^\tau u(\tau)u(s)\exp[-i(\tau - s)]ds\, d\tau\right\}, \quad (28)$$

$$\beta(t) = \frac{1}{\sqrt{2}}\int_0^t u(s)e^{-is}ds, \quad (29)$$

$$\beta^*(t) = \frac{1}{\sqrt{2}}\int_0^t u(s)e^{is}ds. \quad (30)$$

Let us find $|a_{00}(T)|^2$. We denote the expression in curly brackets in (28) by $I(T)$ and transform it. In view of (29) and (30) we have

$$I(T) = -\frac{1}{2}\int_0^T\int_0^t u(t)u(s)\exp[-i(t-s)]ds\, dt =$$

$$= -\frac{1}{2}\int_0^T u(t)e^{-it}\left[\int_0^t u(s)e^{is}ds\right]dt =$$

$$= -\int_0^T\left[\frac{1}{\sqrt{2}}\int_0^t u(s)e^{is}ds\right]d\left[\frac{1}{\sqrt{2}}\int_0^t u(\tau)e^{-i\tau}d\tau\right] = -\int_0^T \beta^*(t)d\beta(t). \quad (31)$$

Using the equalities

$$d|\beta(t)|^2 = d\beta(t)d\beta^*(t) + \beta^*(t)d\beta(t), \quad (32)$$

$$\mathrm{Re}[\beta(t)d\beta^*(t)] = \mathrm{Re}[\beta^*(t)d\beta(t)], \quad (33)$$

we find that $\mathrm{Re}\, I(T) = -\frac{1}{2}|\beta(T)|^2$. Therefore

$$|a_{00}(T)|^2 = |\exp I(T)|^2 = \exp[-2\,\mathrm{Re}\, I(T)] = \exp(-|\beta(t)|^2).$$

Now let us calculate, for instance, the quantity $|a_{n0}(T)|^2$. It follows from (26) that

$$a_{n0}(T) = \frac{a_{00}(T)}{\sqrt{n!}}[i\beta^*(T)]^n. \quad (34)$$

Therefore, the probability that the system will be transferred from the state a_{00} to the state a_{n0} equals

$$|a_{n0}(T)|^2 = |a_{00}(T)|^2 |\beta(T)|^{2n} = \frac{[|\beta(T)|^2]^n}{n!} \exp[-|\beta(T)|^2]. \qquad (35)$$

It is clear that (35) is the Poisson distribution of a discrete random variable with the parameter $y = |\beta(T)|^2$. Therefore

$$|a_{n0}(T)|^2 = y^n e^{-y}. \qquad (36)$$

Now setting $n = n_0$, let us find the value y_0 of the parameter y that maximizes the probability (36). Differentiating the right hand side of (36) with respect to y and setting it equal to zero, we obtain

$$n y^{n-1} e^{-y} - y^n e^{-y} = 0. \qquad (37)$$

The solution of interest to us is

$$y_0 = n_0 = |\beta(T)|^2. \qquad (38)$$

Now we can find the probability maximum

$$\max |a_{n0}(T)| = a_{n_0 0}(T) = \frac{n_0^{n_0} e^{-n_0}}{n_0!} \approx \frac{1}{\sqrt{2\pi n_0}}. \qquad (39)$$

We have used the Stirling's formula

$$m! \approx m^m e^{-m} \sqrt{2\pi m} \qquad (40)$$

for the estimate. Substituting (29) into (38), we obtain the condition for the desired control $u(t)$:

$$\left[\int_0^T u(t) \cos t \, dt\right]^2 + \left[\int_0^T u(t) \sin t \, dt\right]^2 = 2n_0. \qquad (41)$$

Then substituting (38) into (36), we obtain

$$|a_{n0}(T)|^2 = \frac{n_0^n}{n!} e^{-n_0}. \qquad (42)$$

Therefore the force $u(t)$, $0 \le t \le T$, defined by the condition (41), carries out the transition of the one-dimensional oscillator from the zero stationary state into an arbitrary nth stationary state with the maximum probability defined by (39). The desired probability that the oscillator will transfer to an arbitrary nth state is defined by (42). For instance, the maximum probability of transferring from the zero stationary state to the first one equals $w_{10}(T) = |a_{10}(T)|^2 = e^{-1} \approx 0.37$. The maximum probability of transition will be achieved if the control $u(t)$ satisfies the condition

$$\left| \int_0^T u(t) e^{-it} dt \right| = \sqrt{2}.$$

In order to meet this condition, it is sufficient, for instance, to apply the impulse control $u(t) = \sqrt{2}\delta(t)$.

2.9. Control of a One-Dimensional Quantum Oscillator by Varying its Eigenfrequency

Another interesting case of the problem of controlling a quantum oscillator is the problem of transferring it to a new stationary state by varying its frequency or mass. The latter problem can be reduced by means of the substitution (Baz et al., 1971)

$$t' = \int \frac{dt}{m(t)}, \quad \omega' = m\omega$$

to the case where m =const but still with variable frequency. The Schrödinger equation for an oscillator with variable frequency in the absence of an external field has the form

$$i\frac{\partial \psi}{\partial t} = -\frac{1}{2}\frac{\partial^2 \psi}{\partial x^2} + \frac{1}{2}\omega^2(t)x^2\psi. \tag{1}$$

Suppose the frequency $\omega(t)$ is constant prior to $t = 0$ and equals ω_0. Let us assume that the variation of frequency occurs on the interval $[0, T]$. After the instant $t = T$, the frequency generally acquires a new constant value $\omega = \omega_T$. In other words, we suppose the following:

$$\omega(t) = \begin{cases} \omega_0, & t \leq 0, \\ \omega_T, & t \geq T. \end{cases} \tag{2}$$

Strictly speaking, it is supposed that the frequency varies gently in the absence of control, that $\omega(t) \to \omega_0$ as $t \to -\infty$ and that $\omega(t) \to \omega_T$ as $t \to \infty$, that is, for any $\epsilon > 0$ there exist t_0 and t_1 such that $|\omega(t) - \omega_0| < \epsilon$ and

$|\omega(t) - \omega_T| < \epsilon$ as soon as $t < t_0$ and $t > t_1$ respectively. We assume that $t_0 = 0$ and $t_1 = T$.

Under the action of the varying frequency, the oscillator can be transferred to a new energy state during the interval $[0, T]$. We have to remark that the system does not change its quantum state when $\omega = $ const or when $\omega(t)$ is slowly varying (Baz et al., 1971).

Let us consider the following control problem: find $\omega(t)$ such that under its action the system transfers with the maximum (or preassigned) probability from an initial state $|n, \omega_0\rangle$ to a given final state $|m, \omega_T\rangle$ during a finite time T. The probability of this transition is given in Baz et al., 1971:

$$w_{mn} = \frac{n_1!}{n_2!}\sqrt{1-\rho}\,\big|P_{|m+n|/2}^{|m-n|/2}(\sqrt{1-\rho})\big|^2. \tag{3}$$

Here $n_1 = \min(m, n)$, $n_2 = \max(m, n)$, and the $P_j^i(x)$ are the associated Legendre polynomials. The parameter ρ, $0 \le \rho \le 1$, in (3) depends only on the form of $\omega(t)$. It is also shown in Baz et al. (1971) that there is a direct relation between the state change in the given oscillator and the motion of the corresponding classical oscillator. Indeed, following Baz et al., (1971), we can write the equation of motion for the classical oscillator with variable frequency in the form

$$\ddot{\xi} + \omega^2(t)\xi = 0 \tag{4}$$

with the initial condition

$$\xi(t) \sim e^{i(\omega_0 t + \delta)} \quad \text{for } t \le 0. \tag{5}$$

At $t \ge T$, when the frequency becomes $\omega_T =$const, the structure of the solution of equation (4) is very simple (Baz et al., 1971):

$$\xi(t) = C_1 e^{i\omega_T t} - C_2 e^{-i\omega_T t}, \quad |C_1|^2 - |C_2|^2 = \frac{\omega_0}{\omega_T} \equiv \sigma^2. \tag{6}$$

The absolute values of the constants C_1 and C_2, and therefore the state of the system at $t = T$ are completely determined by the parameter ρ present in (3) (Baz et al., 1971):

$$C_1 = e^{i\delta}\sigma\frac{1}{\sqrt{1-\rho}}, \quad C_2 = e^{i\delta}\sigma\frac{\sqrt{\rho}}{\sqrt{1-\rho}}. \tag{7}$$

Let us choose the value of $\rho = \rho_0$ such that the probability of transition from the state $|n, \omega_0\rangle$ into $|m, \omega_T\rangle$ attains its maximum. Then the frequency

$w(t)$ can be found from the solution of the following problem of controlling a classical oscillator: find $w(t)$ transferring the system (4) from the intial state

$$\xi(0) = \xi_0 = \cos\delta, \quad \dot\xi(0) = \xi_1 = -w_0\sin\delta \qquad (8)$$

during time T to the state

$$\xi(T) = \xi_{0*} = \frac{\sigma}{\sqrt{1-\rho_0}}\Big(\cos(w_T T + \delta) - \sqrt{\rho_0}\cos(w_T T - \delta)\Big)$$
$$\dot\xi(T) = \xi_{1*} = -\frac{w_T\sigma}{\sqrt{1-\rho_0}}\Big(\sin(w_T T + \delta) - \sqrt{\rho_0}\sin(w_0 T - \delta)\Big). \qquad (9)$$

Inasmuch as we are interested in the behaviour of the system during the interval $[0, T]$, let us assume that the control is identically zero outside this interval, so we shall henceforth regard this problem as one of finite control.

Now let us solve this problem. In equation (4) we set

$$-w^2(t)\xi \equiv v(t). \qquad (10)$$

Then we obtain the autonomous system in $v(t)$

$$\ddot\xi = v(t) \qquad (11)$$

with the initial conditions (8) and final conditions (9). The function $v(t)$ that solves this problem of control can be written (Appendix 2) in the form

$$v_\Phi(t) = -(\xi_0 - \xi_{0*} + \xi_{1*}T)\delta'(t) - (\xi_1 - \xi_{1*})\delta(t) - \phi''(t), \qquad (12)$$

where $\phi(t)$ is an arbitrary finite function with support $[0, T]$.

The trajectory that corresponds to this control can be determined from the solution of the Cauchy problem

$$\ddot\xi = v(t), \qquad (13)$$

$$\xi(0) = \xi_0, \quad \dot\xi(0) = \xi_1 \qquad (14)$$

and has the following form:

$$\xi_\Phi(t) = \xi_{0*} + \xi_{1*}(t - T) - \int_0^t \phi''(\lambda)(t - \lambda)d\lambda. \qquad (15)$$

But it now follows trivially from (10) that the required control $\omega^2(t)$ is

$$\omega^2(t) = \frac{(\xi_1 - \xi_{1*})\delta(t) + (\xi_0 - \xi_{0*} + \xi_{1*}T)\delta'(t) + \phi''(t)}{\xi_{0*} + \xi_{1*}(t - T) - \int_0^t \phi''(\lambda)(t - \lambda)d\lambda}. \tag{16}$$

The function (16), which solves the control problem for the classical oscillator, is at the same time the solution of the quantum oscillator control problem posed above.

It is evident that $0 < \omega^2(t) < \infty$. But then the right hand side of equation (16) must satisfy the same inequalities. Hence it is easy to find concrete conditions of controllability for the quantum system in question.

Finally, let us take the case where the control action is a vector-valued function $(u(t), \omega(t))$. As above, the problem consists in finding a control $(u(t), \omega(t))$ such that the quantum oscillator transfers from an initial stationary state during time T to a prescribed final state with either maximum or preassigned probability. The transition probability is now (Baz et al, 1971)

$$W_{mn} = \frac{\sqrt{1 - \rho}}{m!n!}|H_{mn}(y_1, y_2)|^2 \exp\{-\nu(1 - \sqrt{\rho}\cos 2\phi)\}, \tag{17}$$

where the $H_{mn}(y_1, y_2)$ are the Hermite polynomials,

$$y_1 = \sqrt{\nu(1 - \rho)}e^{i\phi}, \ y_2 = -\sqrt{\nu}(e^{-i\phi} - \sqrt{\rho}e^{i\phi}), \ \phi = \delta - \beta, \tag{18}$$

and δ and β are the phases of C_1 and C_2 respectively.

Evidently the expressions (8.19) and (16) found above give the form of the desired control $(u(t), \omega(t))$ if we replace ω_0 by ω_T in (8.19) and choose the value (ν_0, ρ_0) so that (17) attains either its maximum or the preassigned value.

2.10. Obtaining a Specified Probability of a Given State of a Charged Particle by Means of an External Magnetic Field

Let us consider one more case of a quantum system with a non-stationary quadratic Hamiltonian where exact solution of the control problem is possible. Assume that a charged particle is in a varying electromagnetic field specified by its potentials

$$A(t) = [H(t) \times \frac{r}{2}], \ \phi(t) = e\chi(t)(x^2 + y^2)/2M, \tag{1}$$

where $H(t)$ is a homogeneous magnetic field varying with time and directed along the z-axis, r is a vector indicating the coordinates of the particle, ϕ is a scalar potential corresponding to the homogeneous distribution of the charge density (the charge density equals $-e\chi/(2\pi Mc^2)$), $\chi(t)$ is an arbitrary function of t, M is the mass of the particle and e is its charge. Such potentials accommodate the classical equations and can be created by appropriate distributions of current and charge densities. For instance, they describe the electromagnetic field of a coil that carries a varying current. The accuracy of description is quite acceptable for many problems.

Suppose that the variation of the electromagnetic field occurs on a finite time interval, that is, $t \in [0, T]$. Beyond this interval, that is, at $t < 0$ and $t > T$, the magnetic field possesses some constant but generally unequal values. The potential $\phi(t)$ is a finite function with support $[0, T]$, that is, with respect to $H(t)$ and $\phi(t)$ we assume that

$$
\begin{aligned}
H(t) = H_{\text{in}}, \quad &\phi(t) = 0, \ t < 0, \\
H(t) = H_{\text{f}}, \quad &\phi(t) = 0, \ t > T.
\end{aligned}
\tag{2}
$$

Then as $t \to \pm\infty$ there exist coherent states and Landau energy levels (Lyubarskii, 1958). The Hamiltonian of such a system disregarding spin phenomena has the form

$$
H = (2M)^{-1}[(P_x - eA_x)^2 + (P_y - eA_y)^2] + e\phi \ (\hbar = c = 1).
\tag{3}
$$

Let us assume that we can vary the field arbitrarily on $[0, T]$.

We pose the following control problem for the quantum system in question: among the admissible controls $H(t)$, that is, consistent with (2), find $H_0(t)$ such that transfers the system defined by the Hamiltonian (3) from an initial state $|\text{in}\rangle$ to a desired final state $|f\rangle$ during time T, the probability of the event being maximum or a preassigned value. In other words, find a control $H_0(t)$ such that the functional

$$
|a(T)|^2 = a^*(T)a(T)
\tag{4}
$$

attains the maximum or preassigned value.

The amplitude $a(T)$ of transition from the initial state $|\text{in}\rangle$ into the final one $|f\rangle$ is given by a matrix element $\langle f|t \to \infty\rangle$, where $|t \to \infty\rangle$ is the limit of the state $|t\rangle$ as $t \to \infty$. The state $|t\rangle$ is the evolution of the initial state $|\text{in}\rangle$ in time when $t > 0$. The initial state corresponds to $t = 0$.

Lyubarskii (1958) gives the explicit forms of coherent states, eigenstates of
the energy operators, projections of the angular momentum on the direction of
the magnetic field, the Green's functions, and the amplitudes and probabilities
of transition between the initial and final coherent states, as well as between the
Landau levels. It is essential that the coherent states are shown in Lyubarskii
(1958) as the eigenstates of the motion integrals

$$A(t) = \frac{e^{-1/2}}{2}[\epsilon(t)(p_x + ip_y) - iM\dot{\epsilon}(t)(y - ix)]\exp\left[\frac{i}{2}\int_0^t \omega(\tau)d\tau\right],$$

$$B(t) = \frac{e^{-1/2}}{2}[\epsilon(t)(p_y + ip_x) - iM\dot{\epsilon}(t)(x - iy)]\exp\left[-\frac{i}{2}\int_0^t \omega(\tau)d\tau\right]. \tag{5}$$

It is also shown that any of these variables can be expressed in terms of one
function $\epsilon(t)$, so that the system under consideration is completely described by
this function. At the same time, $\epsilon(t)$ obeys the classical equation for oscillations
with variable frequency

$$\ddot{\epsilon}(t) + \Omega^2(t)\epsilon(t) = 0, \tag{6}$$

where

$$\Omega^2(t) = \frac{1}{4}\omega^2(t) + \frac{e^2}{M^2}\chi(t), \tag{7}$$

$$\omega^2(t) = \frac{e}{M}H(t). \tag{8}$$

As in Lyubarskii (1958), we shall discuss only the following special solution
of equation (6):

$$\epsilon(t) = |\epsilon|\exp\left(i\frac{e}{M}\int_0^t |\epsilon(\tau)|^{-2}d\tau\right). \tag{9}$$

In compliance with Lyubarskii (1958) we set the initial conditions in the form

$$\epsilon(0) = \left(\frac{2}{H_{in}}\right)^{1/2}e^{i\nu_1}, \quad \dot{\epsilon}(0) = i\left(\frac{1}{2H_{in}}\right)^{1/2}\omega_{in}e^{i\nu_1}, \tag{10}$$

that is, the solution of equation (6) is described in the limit as $t \to \infty$ by the
expressions

$$\epsilon(-\infty) = \left(\frac{2}{H_{in}}\right)^{1/2}\exp\left(\frac{i}{2}\omega_{in}t\right), \quad \dot{\epsilon}(-\infty) = \frac{i}{2}\omega_{in}\epsilon(-\infty). \tag{11}$$

This makes it possible to offer a description of the quantum system in which
the coherent states $|\alpha, \beta; t \to -\infty\rangle$ coincide with the initial coherent states
$|\alpha, \beta; in\rangle$, while the eigenstates of the energy operators and the projections

of the angular momentum on the direction of the magnetic field coincide as $t \rightarrow \infty$ with the Landau solutions for a constant magnetic field H_{in} (Landau and Lifshits, 1963).

The general solution of equation (6) for large enough $t \geq T$ can be presented in the following form (Lyubarskii, 1958):

$$\epsilon_f = \left(\frac{2}{H_f}\right)^{1/2} \xi \exp\left(\frac{1}{2} i \omega_f t\right) - i \left(\frac{2}{H_f}\right)^{1/2} \eta \exp\left(-\frac{1}{2} i \omega_f t\right), \tag{12}$$

where the complex parameters ξ and η are related by the condition

$$|\xi|^2 - |\eta|^2 = 1 \tag{13}$$

and they are completely defined by the values of the transition amplitudes. We give the explicit form of these amplitudes. The amplitude of transition between coherent states $|\alpha, \beta; \text{in}\rangle$ and $|\gamma, \delta; f\rangle$ has the following form (Lyubarskii, 1958):

$$T_{\alpha\beta}^{\gamma\delta} = \langle \gamma, \delta; f | \alpha, \beta; t \rightarrow \infty \rangle = \xi^{-1} \exp\left[-\frac{1}{2}(|\alpha|^2 + |\beta|^2 + \right.$$
$$\left. + |\gamma|^2 + |\delta|^2) + \xi^{-1}(\alpha\beta\eta^* + \beta\delta^* + \alpha\gamma^* - \eta\gamma^*\delta^*)\right] \tag{14}$$

and is the generating function for all the other amplitudes. This function allows us to obtain the amplitude $T_{\alpha\beta}^{m_1 m_2}$ of the transition between the initial coherent states and the final states with a given energy and projection of the angular momentum on the direction of the magnetic field:

$$T_{\alpha\beta}^{m_1 m_2} = \left(\frac{m_2!}{m_1!}\right)^{1/2} \left(-\frac{\eta}{\xi}\right)^{m_2} \left(\frac{\alpha}{\xi}\right)^{m_1-m_2} T_{\alpha\beta}^{00} L_{m_2}^{m_1-m_2}\left(\frac{\alpha\beta}{\xi\eta}\right), \quad m_1 \geq m_2$$
$$T_{\alpha\beta}^{m_1 m_2} = \left(\frac{m_1!}{m_2!}\right)^{1/2} \left(-\frac{\eta}{\xi}\right)^{m_1} \left(\frac{\beta}{\xi}\right)^{m_2-m_1} T_{\alpha\beta}^{00} L_{m_1}^{m_2-m_1}\left(\frac{\alpha\beta}{\xi\eta}\right), \quad m_1 < m_2 \tag{15}$$

where $L_p^s(x)$ are the Laguerre polynomials. Similarly,

$$T_{n_1 n_2}^{\alpha\beta} = \left(\frac{n_2!}{n_1!}\right)^{1/2} \left(-\frac{\eta^*}{\xi}\right)^{n_2} \left(\frac{\gamma^*}{\xi}\right)^{n_1-n_2} T_{00}^{\gamma\delta} L_{n_2}^{n_1-n_2}\left(-\frac{\gamma^*\delta^*}{\xi\eta^*}\right), \quad n_1 \geq n_2,$$
$$T_{n_1 n_2}^{\alpha\beta} = \left(\frac{n_1!}{n_2!}\right)^{1/2} \left(-\frac{\eta^*}{\xi}\right)^{n_1} \left(\frac{\delta^*}{\xi}\right)^{n_2-n_1} T_{00}^{\gamma\delta} L_{n_1}^{n_2-n_1}\left(-\frac{\gamma^*\delta^*}{\xi\eta^*}\right), \quad n_1 < n_2. \tag{16}$$

In the general set-up of control problems, of interest are the amplitudes and probabilities of transitions between the states $|m_1, m_2; \text{in}\rangle$ and $|n_1, n_2; f\rangle$ with fixed energy and projection of the angular momentum on the direction of the magnetic field. If the projection L_z is positive, then these variables are (Lyubarskii, 1958):

a) $\quad T_{n_1 n_2}^{m_1 m_2} = \xi^{-1}(-1)^{m_1-n_1}\left(\dfrac{n_1!m_2!}{n_2!m_1!}\right)^{1/2}\dfrac{(\xi^*)^{n_1}(\eta^*)^{m_1-n_1}}{\xi^{m_2}} \times$

$$\times P_{n_1}^{(m_1-n_1,n_2-n_1)}\left(1-2\left|\frac{\eta^2}{\xi^2}\right|\right),$$

$$n_2 \geq n_1, \ m_i \geq n_i, \ i = 1,2; \qquad (17)$$

b) $\quad T_{n_1 n_2}^{m_1 m_2} = \xi^{-1}\left(\dfrac{n_2!m_1!}{n_1!m_2!}\right)^{1/2}\dfrac{(\xi^*)^{m_1}(\eta^*)^{n_1-m_1}}{\xi^{n_2}} \times$

$$\times P_{m_1}^{(n_1-m_1,n_2-n_1)}\left(1-2\left|\frac{\eta^2}{\xi^2}\right|\right),$$

$$m_2 \geq n_1, \ m_i < n_i, \ i = 1,2. \qquad (18)$$

Here $P_n^{(\alpha,\beta)}$ are the Jacobi polynomials. An interchange of the subscripts $(1 \leftrightarrows 2)$ must be performed in (17) and (18) if $L_z < 0$.

The transition probabilities are represented by the expressions (Lyubarskii, 1958):

$$W_{n_1 n_2}^{m_1 m_2} = \frac{m_2!m_1!}{m_1!n_2!}\, R^{m_1-n_1}(1-R)^{n_2-n_1+1}\left|P_{n_1}^{(m_1-n_1,n_2-n_1)}\left(1-2\left|\frac{\eta^2}{\xi^2}\right|\right)\right|^2, \quad (19)$$

where $R = |\eta/\xi|^2 < 1$. For case b) in (19) we have to interchange $n_i \leftrightarrows m_i$.

The fact that the pairs of parameters ξ and η are determined by the values of the transition amplitudes allows us to approach uniformly the whole set of problems concerning control of the quantum system in question. Indeed, if we use (19) as the functional (4), the control problem posed above can be made concrete as follows: among the admissible controls $H(t)$ find a control $H_0(t)$ such that the system transfers with the maximum (or preassigned) probability from the initial state $|m_1, m_2; \text{in}\rangle$ to a prescribed final state $|n_1, n_2; f\rangle$ during an interval $T > t_0$ or, what is the same, find $H_0(t)$ such that the values of the energy and the projection of the angular momentum on the direction of the magnetic field assume the prescribed eigenvalues k_*, l_* with maximum (or preassigned) probability during an interval $T > t_0$.

Furthermore, we can pose the control problem for coherent states of a charged particle moving in a magnetic field, that is, the problem of finding a control steering the system from the initial coherent state $|\alpha, \beta; \text{in}\rangle$ to the desired final coherent state $|\gamma, \delta; f\rangle$ during time T with the maximum (or pre-assigned) probability. Inasmuch as coherent states are the eigenstates of the integrals of motion (5), the objective of control can be formulated as follows: find the greatest possible probability to reach the state where the operators $A(t)$ and $B(t)$ would acquire certain desired values α_*, β_*. The physical meaning of such a problem consists in that the eigenvalues α, β define (as $t \to \pm\infty$) the coordinates of the orbital centre in the plane (x, y) and the current coordinates of the wave packet centre. The functional of the form (4) can be written in this case if we know the amplitudes (14) of transition between coherent states.

Next we consider a control problem whose objective is the transition from a coherent state to a state with a prescribed fixed energy and projection of the angular momentum on the magnetic field. The functional (4) can be written in explicit form with the use of (15) and (16). We can also consider the problem of transition from the initial 'semicoherent' state $|n_1, \beta; \text{in}\rangle$ to the desired final state $|m_1, \delta; f\rangle$. These 'semicoherent' states are the best classical approximations to the stationary state with fixed energy. Similar formulations of other control problems are also possible.

It is essential for all these problems that there be the possibility of formulating a unified auxiliary problem of controlling a classical oscillator with variable frequency, whose solution is the general solution of each of the above-mentioned problems of controlling quantum systems. These are distinguished by the choice of the quality functional, which is only related to the parameters ξ and η_0. When these parameters are determined from concrete requirements on the functional (4), we can directly pose and solve the auxiliary control problem for the classical oscillator.

In view of (9), this problem consists in the following: find a control $\Omega_0(t) \in \Omega$ that steers the system

$$\frac{d^2|\epsilon|}{dt^2} + \Omega^2(t)|\epsilon| - \left(\frac{e}{M}\right)^2 |\epsilon|^{-3} = 0 \tag{20}$$

from the initial state

$$\epsilon_0 = |\epsilon(0)| = \left(\frac{2}{H_{\text{in}}}\right)^{1/2}, \quad \epsilon_1 = |\dot{\epsilon}(0)| = \left(\frac{\omega_{\text{in}}^2}{2H_{\text{in}}}\right)^{1/2} \tag{21}$$

to the prescribed final state

$$
\begin{aligned}
\epsilon_{*0} &\equiv |\epsilon_f| = \left(\frac{2}{H_f}\right)^{1/2} \{|\xi_0|^2 + |\eta_0|^2 + 2|\xi_0|\,|\eta_0|\cos(\sigma_1 + \sigma_2)\}^{1/2}, \\
\epsilon_{1*} &\equiv |\dot{\epsilon}_f| = \frac{\omega_f}{\sqrt{2H_f}} \{|\xi_0|^2 + |\eta_0|^2 + 2|\xi_0|\,|\eta_0|\cos(\sigma_1 + \sigma_2)\}^{1/2},
\end{aligned}
\tag{22}
$$

where $\xi_0 = |\xi_0|e^{i\sigma_1}$, $\eta_0 = |\eta_0|e^{i\sigma_2}$.

A similar problem of controlling the classical oscillator with variable frequency is considered in detail in Zubov (1975). Therefore without dwelling on the niceties of the solution, we shall cite the form of the function Ω that solves the above problems of controlling the states of a charged particle moving in a variable magnetic field:

$$\Omega_0^2(t) = \left(\frac{e}{M}\right)^2 |\epsilon_\phi|^{-4} - v_\phi(t)|\epsilon_\phi|^{-1}, \tag{23}$$

where

$$v_\phi(t) = -(\epsilon_0 - \epsilon_{0*} + \epsilon_{1*}T)\delta'(t) - (\epsilon_1 - \epsilon_{1*})\delta(t) - \phi''(t), \tag{24}$$

$$|\epsilon_\phi| = \epsilon_{0*} + \epsilon_{1*}(t - T) - \int_0^t \phi''(\lambda)(t - \lambda)d\lambda. \tag{25}$$

Here $\phi(t)$ is an arbitrary finite function with the support $[0, T]$. As in §2.7, the requirements that (23) be positive and bounded define the necessary conditions of controllability.

Another practical example is the problem of controlling a charged particle moving in a constant homogeneous magnetic field when the control is a varying electric field with the components

$$E_x = E_1(t), \ E_y = E_2(t), \ E_z = 0. \tag{26}$$

The control, that is, the electric field, acts within the time interval $[0, T]$. The electric field beyond this interval equals zero.

For simplicity's sake, the direction of the magnetic field is chosen along the z-axis. The Schrödinger equation defining the motion of such a particle in the (x, y) plane has the following form (assuming the mass and charge to be unity) (Lyubarskii, 1958):

$$i\frac{\partial\psi}{\partial t} = -\frac{1}{2}\left(\frac{\partial^2\psi}{\partial x^2} + \frac{\partial^2\psi}{\partial y^2}\right) + \frac{\omega^2}{8}(x^2 + y^2)\psi +$$
$$+ \frac{i\omega}{2}\left(x\frac{\partial\psi}{\partial y} - y\frac{\partial\psi}{\partial x}\right) - (E_1 x + E_2 y)\psi \quad (\hbar = c = 1). \tag{27}$$

Transformation of the variables

$$\xi = -\frac{x + iy}{\sqrt{2}}e^{i\omega t/2}, \ \ \xi^* = \frac{iy - x}{\sqrt{2}}e^{-i\omega t/2}, \tag{28}$$

equivalent to a transition to a moving frame, makes it possible to rewrite equation (27) (Lyubarskii, 1958):

$$i\frac{\partial\psi}{\partial t} = -\frac{\partial^2\psi}{\partial\xi\partial\xi^*} + \frac{\omega^2}{4}|\xi|^2\psi + (u\xi^* + u^*\xi)\psi, \tag{29}$$

where

$$u(t) = \frac{1}{\sqrt{2}}(E_1 + iE_2)e^{i\omega t/2}. \tag{30}$$

The problem of controlling the quantum system consists in finding a function $u(t)$ such that the system transfers from the initial state $|\text{in}\rangle$ to the desired state $|f\rangle$ during time T so that the functional (4) attains a maximum (or preassigned) value.

It is shown in Lyubarskii (1958) that the behaviour of this system can be described completely by a complex variable ξ_0 obeying the classical equation of motion

$$\ddot{\xi}_0 + \frac{\omega^2}{4}\xi_0 = u(t) \tag{31}$$

with the initial conditions

$$\xi_0 = \dot{\xi}_0 = 0. \tag{32}$$

For sufficiently large $t \geq T$, the variable ξ_0 is given as

$$\xi_0 = \frac{1}{\sqrt{\omega}}\Big[\lambda\exp\Big(\frac{i\omega t}{2}\Big) - i\mu\exp\Big(-\frac{i\omega t}{2}\Big)\Big], \tag{33}$$

where the complex parameters λ and μ enter the expressions for the amplitudes of transitions between the coherent states $|\alpha,\beta;\text{in}\rangle$ and $|\gamma,\delta:f\rangle$:

$$\langle\gamma,\delta;f|\alpha,\beta;\text{in}\rangle = \langle 0,0;f|0,0;\text{in}\rangle\exp\Big[-\frac{1}{2}(|\lambda|^2 + |\beta|^2 + |\gamma|^2 + |\delta|^2 +$$

$$+\alpha\gamma^* + \beta v^* - \mu^*\alpha - \lambda\beta - \mu\gamma^* + \lambda^*\delta^*\Big], \tag{34}$$

$$\langle 0,0;f|0,0;\text{in}\rangle = \exp\Big[-\frac{1}{2}(|\lambda|^2 + |\mu|^2) + \frac{i}{2}\int_0^t(u\xi^* + u^*\xi_0)d\tau\Big] \tag{35}$$

as well as between the states with given energy and angular momentum:

$$\langle m_1,m_2;f|n_1,n_2;\text{in}\rangle = \langle 0,0;f|0,0;\text{in}\rangle(m_1!m_2!n_1!n_2!)(q_1!q_2!)^{-1}\times$$

$$\times(-\mu/\mu^*)^{m_1-n_1}|\mu|^{|m_1-n_1|/2}(\lambda^*/\lambda)^{m_2-n_2}|\lambda|^{|m_2-n_2|/2}\times$$

$$\times L_{p_1}^{q_1-p_1}(|\mu|^2)L_{p_2}^{q_2-p_2}(|\sigma|^2), \tag{36}$$

where

$$p_1 = \frac{1}{2}(n_1 + m_1 - |n_1 - m_1|), \quad p_2 = \frac{1}{2}(n_2 + m_2 - |n_2 - m_2|),$$

$$q_1 = \frac{1}{2}(n_1 + m_1 + |n_1 - m_1|), \quad q_2 = \frac{1}{2}(n_2 + m_2 + |n_2 - m_2|),$$

are determined from the requirements on the functional (4). Note that we can also take the amplitude of transition between states to be the functional.

Suppose that $\lambda = \lambda_0$ and $\mu = \mu_0$ are the values of the parameters λ and μ at which the functional (4) reaches its maximum (or preassigned) value. Then the control $u_0(t)$ that solves the problem formulated above can be found from the solution of the following auxiliary problem: find a function $u_0(t)$ steering the system (31) and (32) during time T to the desired state

$$\xi_0 = \frac{1}{\sqrt{\omega}}\left[\lambda_0 \exp i\left(\frac{\omega T}{2} + \nu_2\right) - i\mu_0 \exp i\left(-\frac{\omega T}{2} + \nu_2\right)\right] \equiv \xi_{0*},$$
$$\dot{\xi} = i\frac{\sqrt{\omega}}{2}\left[\lambda_0 \exp i\left(\frac{\omega T}{2} + \nu_2\right) + i\mu_0 \exp i\left(-\frac{\omega T}{2} + \nu_2\right)\right] \equiv \dot{\xi}_{0*}. \tag{37}$$

Using the form of the general solution of the finite control problem (Appendix 2) we obtain

$$u_o(t) = -\xi_{0*}\delta'(t) - \dot{\xi}_{0*}\delta(t) + \frac{\omega^2}{4}\phi(t) - \phi''(t), \tag{38}$$

where $\phi(t)$ is an arbitrary finite function on $[0, T]$. Therefore

$$E_1 = \sqrt{2}\exp(-i\omega t/2)\operatorname{Re} u, \quad E_2 = \sqrt{2}\exp(-i\omega t/2)\operatorname{Im} u. \tag{39}$$

Note that for the given quantum system the above general set-up of the control problem can be made more concrete depending on the final objective of control. This only concerns the choice of the functional (4) in its explicit form and therefore concerns the values λ_0 and μ_0. When the objective of control is being chosen, one should take into account the fact that the states $|n_1, n_2; t\rangle$ of this system are also the eigenstates of the operators of energy and angular momentum in the moving frame, while α and β give the initial coordinates of motion in a circle and the coordinates of the centre of this circle respectively. Here the velocity of motion of the reference frame defined by (28) is not constant for a varying electric field.

2.11., Control of the State of a Free Particle by an External Force

We shall deal in this section with the following problem: which new states can a free particle acquire if it is controlled by an external force and if it has a given state at the initial instant ? The Green's function for a free particle has the form (Feynman and Hibbs, 1965)

$$G_1(x,\xi,t) = \left(\frac{m}{2\pi i\hbar T}\right)^{1/2} \exp\left(\frac{i}{\hbar}S_{\mathrm{kl}}\right), \tag{1}$$

where

$$S_{\mathrm{kl}} = \frac{m}{2T}(x-\xi)^2 - \frac{x-\xi}{T}\int_0^T f(\theta)(T-\theta)d\theta +$$

$$+ x\int_0^T f(\theta)d\theta - \frac{1}{mT}\int_0^T\int_0^\theta f(\theta)f(s)(T-\theta)s\,ds\,d\theta.$$

We set

$$A(T) = \left(\frac{m}{2\pi i\hbar T}\right)^{1/2}, \tag{2}$$

$$B(T,x) = x\int_0^T f(\theta)d\theta - \frac{1}{mT}\int_0^T\int_0^\theta f(\theta)f(s)(T-\theta)s\,ds\,d\theta, \tag{3}$$

$$M_1(T) = \int_0^T f(\theta)d\theta, \tag{4}$$

$$M_2(T) - \int_0^T f(\theta)\theta\,d\theta, \tag{5}$$

$$M_3(T) = \int_0^T\int_0^\theta f(\theta)f(s)s\,ds\,d\theta. \tag{6}$$

We shall investigate the system in terms of a new function

$$P(t) = \int_0^t f(\theta)d\theta \tag{7}$$

possessing the physical sense of momentum, rather than in terms of the external force $f(t)$. This may be convenient because when we apply the Fourier transformation we arrive precisely at the momentum representation of the wave function. In terms of the new variables, we obtain the following expressions for the parameters (2)–(6) of the Green's functions:

$$M_1(T) = P(T), \tag{8}$$

$$M_2(T) = \int_0^T P'(\theta)\theta \, d\theta = P(T)T - \int_0^T P(\theta)d\theta, \tag{9}$$

$$\blacksquare \quad M_3(T) = \int_0^T P'(\theta) \int_0^\theta P'(s)s \, ds \, d\theta =$$

$$= \int_0^T P'(s)s \, ds - P(T) - \int_o^T P(\theta)P'(\theta)\theta \, d\theta =$$

$$= P(T)M_2(T) - \frac{P^2(T)T}{2} + \frac{1}{4}\int_0^T P^2(\theta)d\theta,$$

$$\int_0^T \int_0^\theta f(\theta)f(s)(T - \theta)s \, ds \, d\theta = TM_3(T) - \frac{M_2^2(T)}{2}. \quad \square \tag{10}$$

Let us apply to both sides of the equation

$$\psi_*(x,t) = \int_{-\infty}^\infty G(x,\xi,t)\psi_0(\xi)d\xi \tag{11}$$

the Fourier transformation

$$F_x(\phi) = \int_{-\infty}^\infty \phi(x)e^{-i\gamma x}dx, \tag{12}$$

where p is the momentum and $\gamma = p/\hbar$. Equation (11) defines the change of state of a free particle at $t > 0$ if at $t = 0$ it was in the state $\psi(x,0) = \psi_0(x)$. The inverse Fourier transform has the following form:

$$\phi(x) = \int_{-\infty}^\infty \frac{F_x(\phi)e^{i\gamma x}}{2\pi}d\gamma. \tag{13}$$

We convert (11) to a form convenient for our purposes:

$$\psi_*(x) = A(T)\exp\left[\frac{i}{\hbar}xM_1(T)\right]\exp\left\{-\frac{i}{mT^*}\left[TM_3(T) - \frac{M_2^2(T)}{2}\right]\right\} \times$$

$$\times \int_{-\infty}^\infty \exp\left\{\frac{i}{\hbar}\left[\frac{m}{2T}(x-\xi)^2 - (x-\xi)\left(M_1(T) - \frac{M_2(T)}{T}\right)\right]\right\}\psi_0(\xi)d\xi.$$

It follows readily that

$$F_\gamma\left\{\psi_*(x)\exp\left[-\frac{i}{\hbar}xM_1(T)\right]\right\} = \tilde{\psi}_*\left[\gamma + \frac{M_1(T)}{\hbar}\right],$$

$$F_\gamma\left[\int_{-\infty}^\infty\left\{\exp\left[\frac{i}{\hbar}\left[\frac{m}{2T}(x-\xi)^2 - (x-\xi)\left(M_1(T) - \frac{M_2(T)}{T}\right)\right]\right]\right\}\psi_0(\xi)d\xi\right] =$$

$$= \tilde{\psi}_0(\gamma)\frac{1}{A(T)}\exp\left\{-\frac{1}{\hbar}\left[M_1(T) - \frac{M_2(T)}{T} + \hbar\gamma\right]^2 T\right\}. \tag{14}$$

Therefore we arrive at the following relation:

$$\tilde{\psi}_* \left(\gamma + \frac{M_1(T)}{\hbar} \right) = \tilde{\psi}_0(\gamma) \exp\left\{ -\frac{i}{m\hbar} \left[M_3(T) - \frac{M_2^2(T)}{2T} \right] - \right.$$
$$\left. -\frac{T_2}{2} \left[M_1(T) - \frac{M_2}{T} + \hbar\gamma \right]^2 \right\},$$

or

$$\tilde{\psi}_* \left(\gamma + \frac{M_1(T)}{\hbar} \right) = \tilde{\psi}_0(\gamma) \exp\left\{ -\frac{i}{m\hbar} \left[M_2(T)P(T) - \frac{P^2(T)T}{2} + \right. \right.$$
$$+\frac{1}{4} \int_0^T P^2(\theta)d\theta - \frac{M_2^2(T)}{2T} - \frac{TM_1^2(T)}{2} + M_2(T)M_1(T) -$$
$$-\frac{M_2^2(T)}{2T} - T\hbar\gamma \left[M_1(T) - \frac{M_2(T)}{T} \right] - \frac{T\hbar^2\gamma^2}{2} \right]\bigg\} =$$
$$= \tilde{\psi}_0(\gamma) \exp\left\{ -\frac{i}{m\hbar} \left[-G(T) \int_0^T G(\theta)d\theta + \right. \right.$$
$$+\frac{1}{4} \int_0^T G^2(\theta)d\theta - \frac{\left[G(T)T - \int_0^T G(\theta)d\theta \right]^2}{T} +$$
$$+G^2(T)T - G(T) \int_0^T G(\theta)d\theta - \hbar\gamma \int_0^T G(\theta)d\theta - \frac{T\hbar^2\gamma^2}{2} \right]\bigg\} =$$
$$= \tilde{\psi}_0(\gamma) \exp\left\{ -\frac{i}{m\hbar} \left[\frac{1}{4} \int_0^T G^2(\theta)d\theta - \frac{1}{T} \left[\int_0^T G(\theta)d\theta \right]^2 - \right. \right.$$
$$-\hbar\gamma \int_0^T G(\theta)d\theta - \frac{T\hbar^2\gamma^2}{2} \right]\bigg\}. \tag{15}$$

We set

$$G(T) = G_0, \tag{16}$$

$$\int_0^T G(\theta)d\theta = G_1, \tag{17}$$

$$\int_0^T G^2(\theta)d\theta = G_2. \tag{18}$$

Therefore we arrive at the final relation

$$\tilde{\psi}_* \left(\gamma + \frac{G_0}{\hbar} \right) = \tilde{\psi}_0(\gamma) \exp\left[\frac{i}{m\hbar} \left(-\frac{1}{4}G_2 + \frac{G_1^2}{T} + \hbar\gamma G_1 + \frac{T\hbar^2\gamma^2}{2} \right) \right]. \tag{19}$$

Let $|\tilde{\psi}_0|^2 = \phi(\gamma)$, where

$$\int_{-\infty}^{\infty} \phi(\gamma)d\gamma = 1.$$

Then

$$|\tilde{\psi}_*\left(\gamma + \frac{G_0}{\hbar}\right)|^2 = |\tilde{\psi}_0(\gamma)|^2. \tag{20}$$

The inference is that when a free particle is controlled by an external impulse $G(t)$ or a force $f(t)$, we can only arrive from the initial momentum distribution of the free particle to the same distribution shifted by an amount proportional to the impulse $G(T) = \int_0^T f(\theta)d\theta$. This result was obtained by N.L. Lepe.

2.12. Control of the Coefficients of Linear Differential Equations. Impulse Control

It is rather common that control enters the description of a quantum system so that the coefficients of linear differential equations depend on the control. Illustrative examples of such a system were considered above (the Schrödinger equation, the Heisenberg equation, equations in amplitudes in perturbation theory, etc.). Suppose that a controlled quantum system is defined by a system of linear differential equations whose matrix form is

$$\dot{a}(t) = A(t, u(t))a(t), \ a(t_0) = a_0, \ t \geq t_0, \tag{1}$$

where $A(t, u)$ is a quadratic matrix that depends on the control and the vector $a(t)$ defines the state of the system time t. The elements of the matrix A and the vector a are generally complex.

The solution of (1) exists and is unique under rather broad assumptions with respect to the form of dependence of $A(t, u(t))$ on t, the assumptions being sufficient for the solution of practical problems (Coddington and Levinson, 1953; Hartman, 1964). As is known (Coddington and Levinson, 1953; Hartman, 1964), the solution $a(t)$ of the system (1) can be written for each fixed $u(t)$ at $t > t' \geq t_0$ in the following form:

$$a(t) = \Phi(t, t')a(t'), \tag{2}$$

where $a(t')$ is the state of the system at the instant t' and $\Phi(t, t')$ is the fundamental matrix of the system (1) with the property:

$$\Phi(t, t') = I, \ t = t', \tag{3}$$

where I is the identity matrix. The matrix $\Phi(t, t')$ has the following property:

$$|\Phi(t, t')| = \exp \int_{t'}^{t} \mathrm{Sp} A(\tau, u(\tau)) d\tau, \tag{4}$$

where Sp is the trace of the matrix.

Unfortunately, there are no general methods of finding the fundamental matrix $\Phi(t, t')$. But in a particular case, when the system (1) consists of one equation $(n = 1)$, the solution can be found in the following general form:

$$a(t) = a_0 \exp \int_{t_0}^{t} A(\tau, u(\tau)) d\tau. \tag{5}$$

Strictly speaking, this solution cannot be generalized if n is arbitrary. However, for instance, when the matrix A is chronologically commutable, that is, when

$$A(t, u(t)) A(t', u(t')) = A(t', u(t')) A(t, u(t)) \tag{6}$$

for any $t \geq t_0$ and $t' \geq t_0$, the solution of (1) can be written in the form

$$a(t) = \left[\exp \int_{t_0}^{t} A(\tau, u(\tau)) d\tau \right] a_0. \tag{7}$$

For instance, the solution of (1) for the time-independent A takes the form

$$a(t) = [\exp A(t - t_0)] a_0. \tag{8}$$

When the matrices $A(t, u(t))$ for different instants do not commute with each other, the solution of (1) can be written with the aid of the operator T which is the Dyson chronological operator

$$a(t) = \left[T \exp \int_{t_0}^{t} A(\tau, u(t)) d\tau \right] a_0. \tag{9}$$

Generally, the action of this operator can be rather complicated while the result is, in fact, equivalent to the operator (matrix) which is called the *multiplicative integral* (Gantmakher, 1966).

The multiplicative integral F for the matrix $A(t)$ with continuous elements is defined as the limit of the product of matrices

$$F(t, t_0) = \lim_{\Delta t_k \to 0} [\exp A(\tau_n, u(\tau_n)) \Delta t_n \ldots \exp A(\tau_1, u(\tau_1)) \Delta t_1] =$$

$$= \lim_{\Delta t_k \to 0} [(I + A(\tau_n, u(\tau_n)) \Delta t_n \ldots (I + A(\tau_1, u(\tau_1)) \Delta t_1)] =$$

$$= \int_{t_0}^{t} [I + A(\tau, u(\tau)) d\tau], \tag{10}$$

where $t_1, \ldots, t_n$ separate the interval (t_0, t) into n parts $\Delta t_k = t_k - t_{k-1}$ ($k = 1, 2, \ldots, n$, $t_n = t$) and τ_k is a point within the interval (t_{k-1}, t_k).

The solution of (1) can be written in terms of $F(t, t_0)$:

$$a(t) = F(t, t_0)a_0, \quad F(t_0, t_0) = I. \tag{11}$$

Therefore, the multiplicative integral $F(t, t_0)$ coincides with the fundamental matrix $\Phi(t, t_0)$. The expression under the lim sign in (10) can be used for the approximate calculation of the matrix $\Phi(t, t_0)$. Note that the multiplicative integral (10) exists not only for continuous matrices but also under more general assumptions (Coddington and Levinson, 1953).

Now let us consider the case of the matrix $A(t, u(t))$ in (1) having the form

$$A(t, u) = A_0(t, u(t)) + \sum_{j=1}^{m} A_j(u_j)\delta(t - t_j), \tag{12}$$

where $A_0(t, u(t))$ is the matrix of the form considered above, $A_j(u_j)$ is a constant (time-independent) matrix depending only on the control parameter u_j, $j = 1, \ldots, m$ and $t_0 < t_1 < t_2 < \ldots < t_m$. The matrix (12) corresponds to the case when besides (or instead of, when $A_0 = 0$) continuous control $u(t)$ there are impulse controls with the intensity of impulses determined by the matrix $A_j(u_j)$ with control parameters u_j.

The solution of the system (1) with matrix (12) can be written in the form

$$a(t) = \left[\Phi_0(t, t_j) \overset{2}{\underset{j=k}{\otimes}} \exp A_j(u_j)\Phi(t_j, t_{j-1}) \exp A_1(t_1)\right]a_0,$$
$$t_{k+1} > t > t_k, \quad k = 2, \ldots, m, \tag{13}$$

where $\Phi_0(t, t')$ is the fundamental matrix of the system (1) when the matrix $A(t, u(t)) = A_0(t, u(t))$. (The symbol $\otimes$ denotes the chronological arrangement of the product factors.) The system (1) with the matrix (12) is therefore the usual continuous solution between each two neighbouring instants t_j and t_{j+1}. A jump occurs at the instants t_j. For instance, when $A_0(t, u(t)) = 0$ and only the impulse part of the matrix (12) remains, the solution of (1) has the form

$$a(t) = \left[\prod_{j=k}^{m} \exp A_j(u_j)\right]a_0, \quad k = 1, \ldots, m, \; t_{k+1} > t > t_k. \tag{14}$$

The solution (14) is a staircase function with jumps at the instants t_j, $j = 1, 2, \ldots, m$. It is evident that if the pair products of the matrices $A_j(u_j)$ commute, then

$$a(t) = [\exp \sum A_j(u_j)]a_0, \quad k = 1, \ldots, m, \ t_{k+1} > t > t_k.$$

Note that when two-level quantum systems are considered, the researcher often deals with a system of the form (1) consisting of two equations ($n = 2$). Very convenient in this case may be a rather simple formula cited in Lappo-Danilevskii (1957) for the calculation of an arbitrary differential function $f(A)$, where A is a second order matrix:

$$f(A) = \begin{cases} \dfrac{\lambda_2 f(\lambda_1) - \lambda_1 f(\lambda_1)}{\lambda_2 - \lambda_1} I + \dfrac{f(\lambda_2) - f(\lambda_1)}{\lambda_2 - \lambda_1} A, & \text{if } \lambda_1 \neq \lambda_2, \\ [f(\lambda_1) - \lambda_1 f'(\lambda_1)]I + f'(\lambda_1), & \text{if } \lambda_1 = \lambda_2. \end{cases} \quad (15)$$

Here λ_1 and λ_2 are the eigenvalues of the matrix A. If the matrices are of greater dimension, the calculation of the matrix function can be performed by the Lagrange-Sylvester interpolation formulas (Gantmakher, 1966; Lappo-Danilevskii, 1957).

Suppose that the matrix A in (1) has the form

$$A(t, u) \sum_{j=1}^{m} A_j(t, u_j)\delta(t - t_j) = \sum_{j=1}^{m} A(t_j, u_j)\delta(t - t_j), \quad (16)$$

where u_j and t_j, $j = 1, \ldots, m$, are the controlled parameters. In this case, the problem of finite control for steering the system (1) from the initial state $a(t_0) = a_0$ into the desired final state reduces to finding the parameters $u_j, t_j \in [0, T]$, $j = 1, 2, \ldots, m$, and the number m of them such that the following condition holds:

$$a_1 = \exp A_m(t_m, u_m) \exp A_{m-1}(t_{m-1}, u_{m-1}) \ldots \exp A_1(t_1, u_1)a_0. \quad (17)$$

This condition can be obtained from (7). Equation (17) may possess more than one solution u_j, t_j, $j = 1, 2, \ldots, m$. Therefore various problems of optimization can be formulated. For instance, find the minimum integer m for which the system (17) possesses a solution. In another case, it makes sense to choose the solutions of (17) such that a function reaches an extremum or additional conditions of some sort hold.

By way of illustrating the applications of this method, let us consider a finite control problem for a two-level system with the basis states ψ_1 and ψ_2 by the example of controlling the spin of a microparticle by a magnetic field. The amplitudes a_1 and a_2 of these states obey the equations (see §6)

$$\begin{aligned} \dot{a}_1 &= a_1 i\omega \cos\theta + a_2 i\omega e^{-i\phi} \sin\theta, \\ \dot{a}_2 &= a_1 i\omega e^{i\phi} \sin\theta - a_2 i\omega \cos\theta, \end{aligned} \tag{18}$$

where $\omega = \mu|B|/\hbar \geq 0$; μ is the magnetic moment of the particle, B is the vector of the strength of the magnetic field, ϕ and θ are the spherical coordinates of the vector B that begins at the origin. Find a control $\omega(t)$ and the angles ϕ and θ, the angles being time-independent, such that the system (18) transfers from the initial state

$$a_1(0) = \alpha_1, \quad a_2(0) = \alpha_2, $$
$$|\alpha_1|^2 + |\alpha_2|^2 = 1, \quad (\alpha_1\alpha_1^* + \alpha_2\alpha_2^* = 1) \tag{19}$$

to the final state

$$a_1(T) = \beta_1, \quad a_2(T) = \beta_2, $$
$$|\beta_1|^2 + |\beta_2|^2 = 1 \quad (\beta_1\beta_1^* + \beta_2\beta_2^* = 1) \tag{20}$$

during a given time $T \geq 0$.

The matrix of the system (18) is

$$\begin{pmatrix} i\omega \cos\theta & i\omega e^{-i\phi} \sin\theta \\ i\omega e^{i\phi} \sin\theta & -i\omega \cos\theta \end{pmatrix} = i\omega(t)A', \tag{21}$$

where

$$A' = \begin{pmatrix} \cos\theta & e^{-i\phi} \sin\theta \\ e^{i\phi} \sin\theta & -\cos\theta \end{pmatrix} \tag{22}$$

is a constant matrix. It is evident that if t and t' are two different instants, then

$$A(t)A(t') = A(t')A(t), \tag{23}$$

that is, the values of the matrix-valued function $A(t)$ at different instants commute. Therefore we can write the solution of (18) in explicit form (without the use of the Dyson operator):

$$a(t) = \left[\exp \int_0^t A(\tau)d\tau\right] a(0), \quad a(t) = \begin{pmatrix} a_1(t) \\ a_2(t) \end{pmatrix}. \tag{24}$$

The eigenvalues of the matrix $\int_0^t A(\tau)d\tau$ can be found from the equation

$$\begin{vmatrix} iu(t)\cos\theta - \lambda & iu(t)e^{-i\phi}\sin\theta \\ iu(t)e^{i\phi}\sin\theta & -iu(t)\cos\theta - \lambda \end{vmatrix} = 0, \tag{25}$$

where

$$u(t) = \gamma(t) = \int_0^t \omega(\tau)d\tau.$$

It follows from (25) that

$$\gamma^2\cos^2\theta + \lambda^2 + \gamma^2\sin^2\theta = 0,$$

whence

$$\lambda_1 = i\gamma, \quad \lambda_2 = -i\gamma, \tag{26}$$

Using the formula (102) from Lappo-Danilevskii (1957), we can represent the expression in square brackets in (24) in the explicit form

$$\exp\int_0^t A(\tau)d\tau = \frac{-i\gamma(t)e^{i\gamma(t)} - i\gamma(t)e^{-i\gamma(t)}}{-i\gamma(t) - i\gamma(t)}I +$$

$$+ \frac{e^{-i\gamma(t)} - e^{i\gamma(t)}}{-i\gamma(t) - i\gamma(t)}\int_0^t A(\tau)d\tau = \tfrac{1}{2}[e^{i\gamma(t)} + e^{-i\gamma(t)}]I +$$

$$+ \tfrac{1}{2}[e^{i\gamma(t)} - e^{-i\gamma(t)}]A' = I\cos\gamma(t) + iA'\sin\gamma(t), \tag{27}$$

where I is the identity matrix. Therefore the fundamental matrix $\Phi(t)$ of the system (18) has the form

$$\Phi(t) = \exp\int_0^t A(\tau)d\tau = \cos\gamma(t)I + i\sin\gamma(t)A'. \tag{28}$$

It is also obvious that the matrix $\Phi(t)$ is normalized in the sense that

$$\Phi(0) = I. \tag{29}$$

Therefore the solution of (18) subject to the initial condition (19) has the form

$$a(t) = \Phi(t)a(0), \quad \text{or} \quad a(t) = \Phi(t)\alpha,$$

$$\alpha = \begin{pmatrix} \alpha_1 \\ \alpha_2 \end{pmatrix} = a(0) = \begin{pmatrix} a_1(0) \\ a_2(0) \end{pmatrix}. \tag{30}$$

At the instant $t = T$, the condition (20) must hold. Therefore the unknowns $\omega(t)$, ϕ, θ are found from the system of equations

$$\beta = \Phi(T)\alpha. \tag{31}$$

Taking into account the fact that $I\alpha = \alpha$ and

$$A'\alpha = \begin{pmatrix} \alpha_1 \cos\theta + \alpha_2 e^{-i\phi} \sin\theta \\ \alpha_1 e^{i\phi} \sin\theta - \alpha_2 \cos\theta \end{pmatrix},$$

we can obtain from (31) and (28) the explicit form of the system in order to find the unknowns u, ϕ, θ:

$$\begin{aligned}
\beta_1 &= \alpha_1 \cos\gamma + i\alpha_1 \sin\gamma \cos\theta + i\alpha_2 \sin\gamma e^{-i\phi} \sin\theta, \\
\beta_2 &= \alpha_2 \cos\gamma + i\alpha_1 \sin\gamma e^{i\phi} \sin\theta - i\alpha_2 \sin\gamma \cos\theta.
\end{aligned} \tag{32}$$

Setting

$$u = \cos\gamma + i\sin\gamma\cos\theta, \quad v = i\sin\gamma\sin\theta e^{-i\phi} \tag{33}$$

and taking into consideration the fact that the unknown values are real numbers, we arrive at the system

$$\beta_1 = \alpha_1 u + \alpha_2 v, \quad \beta_2 = \alpha_2 u^* - \alpha_1 v^*. \tag{34}$$

We replace the second equation in (34) by its conjugate. This yields the system of two linear algebraic equations in u and v:

$$\beta_1 = \alpha_1 u + \alpha_2 v, \quad \beta_2^* = \alpha_2^* u - \alpha_1^* v. \tag{35}$$

It is easy to see that this always has a unique solution (u, v). Substituting this solution into (33), we arrive at a system of two transcendental equations in γ, ϕ, θ that can be solved easily.

Thus we arrive again at the same result that was obtained in §2.6.

2.13. Control of Magnetization

Let us consider the process of varying the magnetization that is described by the magnetization vector M or in other words, the magnetic moment vector. In the simplest case, when the sole cause of its variation is an external magnetic field (in particular, when the relaxation effects are negligible), the equation describing this process is called the *Bloch equation* which has the form

$$\dot{M} = \gamma[M, H], \tag{1}$$

where γ is a positive scalar quantity and $[M, H]$ is the vector product.

All vectors will be considered here to be polar, emanating from the point 0. In particular, the Bloch equations (1) can be obtained if the spin variables in the quantum equations are averaged. Taking the scalar product of each side of (1) with M, we see that

$$M^2 = \text{const.} \tag{2}$$

Equation (2) is an expression of the law of conservation of the absolute value of the vector M and the fact that as it moves, the end of the vector M is always on the surface of a sphere with centre at the point 0. The radius ρ of this sphere is determined by the initial value of the vector M, that is,

$$\rho = |M(0)| = |M_0| = |M(t)|. \tag{3}$$

Now if the direction of the vector H does not vary (while its absolute value may vary with time), then taking the scalar product of each side of (1) and H yields

$$M \frac{H}{|H|} = \text{const.} \tag{4}$$

This means that when H does not vary in its direction, the angle between M and H remains constant and is, of course, equal to the initial angle between M_0 and H.

Therefore if the direction of the vector H does not vary, there occurs a precession of the vector M about the vector H. The vector M remains stable in its direction if M and H are collinear.

It is also easy to show (see, for instance, Vonsovskii, 1971; Gurevich, 1973) that this precession occurs with an angular velocity whose vector ω is determined by the equation

$$\omega = -\gamma H. \tag{5}$$

As is clear from (5), the vector M rotates clockwise if it is viewed from the end of the vector H.

Now let us consider the problem of finite control of the vector M, which consists in finding a vector-valued function H such that the system (1) would transfer under the action of H from a given initial state $M(0) = M_0$ to a prescribed final state $M(T) = M_*$ during time $T > 0$. Naturally, by virtue of (2) and (3), we have

$$|M_0| = |M_*| = \rho. \tag{6}$$

Obviously the solution of this problem is not unique. However, one solution can be found very easily on the basis of the above considerations. It is evident that if M_0 and M_* are not collinear, it is sufficient to assume that

$$H(k) = k(t)[M_0, M_*], \tag{7}$$

where $k(t)$ is an arbitrary scalar function. ,Indeed, in this case the vector M performs a precession around the vector H while moving from the initial position M_0 to the desired final position M_*, the end of M remaining in the plane of a large circle passing through the ends of M_0 and M_*.

It follows from (5) that the time T of finite control can be determined from the condition

$$\phi = \int_0^T |\omega| dt = \int_0^T \gamma|H| dt = \int_0^T \gamma k(t)|[M_0, M_*]| dt = \int_0^T \gamma k(t)\rho^2 \sin\phi \, dt, \tag{8}$$

where ϕ is the angle between the vectors M_0 and M_* $(0 < \phi < \pi)$. If $k(t) = k = $const, then we find from (8) that $\phi = Tk\rho^2\gamma\sin\phi$, whence

$$T = \frac{\phi}{\gamma k\rho^2\sin\phi}. \tag{9}$$

Here the time T is the minimum time of the process of transition from M_0 into M_* because the transition occurs along the shortest angular path.

If M_0 and M_* are collinear, it is only natural to consider $M_0 = -M_*$, and it is sufficient to take as the required vector an arbitrary vector $H(t)$ at right angles to M_0:

$$H(t) = k(t)h, \quad hM_0 = 0, \quad |h| = 1. \tag{10}$$

The time T of finite control can be determined in this case from the equality

$$\pi = \int_0^T |\omega| dt = \int_0^T \gamma H(t) dt = \int_0^T \gamma k(t) dt. \tag{11}$$

For instance, if $k(t) = k = $ const, it follows that

$$T = \frac{\pi}{\gamma k}. \tag{12}$$

It is clear from conditions (8) and (11) that the time of finite control (of the process of transition) can be made arbitrarily small if in (7) and (10) sufficiently large absolute values of the control vector $H(t)$ are admissible. In the limit, if we assume that $k(t) = k_1 \delta(t)$, where k_1 is a constant and $\delta(t)$ is a delta-function, then the transition from M_0 to M_* occurs instantaneously. At the same time, if M_0 and M_* are non-collinear, then

$$k_1 = \frac{\phi}{\gamma \rho^2 \sin \phi}, \tag{13}$$

and if $M_0 = -M_*$ we have

$$k_1 = \frac{\pi}{\gamma}. \tag{14}$$

Note that the conditions (8) and (11) are similar to the 'theorem of areas' widely used in the theory of dynamic resonance in quantum systems (Loudon, 1973; Macomber, 1976).

In compliance with conditions occurring in practice, this problem of finite control can be made more complicated in the sense that we can take one or even two coordinates of the control vector H, say, H_2 and H_3 as fixed and at least one of these coordinates non-zero, and only one coordinate, for example, $H_1(t)$ varying with time. It is shown in §A2.3 that this finite control problem also has a solution, although it is not so simple as in the case of freely varying vector $H(t)$.

CHAPTER 3

Controllability and Finite Control (Algebraic Methods)

3.1. Algebraic Conditions for the Controllability of a Quantum Process

In order to apply the techniques developed by Brockett (1973a, 1973b) and by Jurdjevič and Sussman (1972), it is worth while considering the control of a quantum system as a set of processes on the Lie group $\mathcal{G}$ of all unitary transformations of the sphere of admissible states into itself. The group $\mathcal{G}$ is associated with its Lie algebra $L(\mathcal{G})$ defined by Hermitian operators in the space of the system's states, and multiplication is defined in $L(\mathcal{G})$ as operator commutation.

The set of accessible states defines the domain of accessibility with respect to an initial state. Domains of accessibility are interrelated through group transformations acting upon appropriate initial states. When the domain of accessibility is the sphere of all possible states in a given space, the system is regarded as *completely controllable*, or simply *controllable*. The domain of accessibility from an arbitrary initial state ψ_0 is associated with a set $\mathcal{A}_U \subset \mathcal{G}$ of admissible controlled transformations $\widehat{V}$ that are accessible from the identity transformation and correspond to a class $\mathcal{U}$ of admissible controls; therefore $\widehat{V} \in \mathcal{A}_U$ if and only if $V\psi_0$ is accessible from $\widehat{T} \in \mathcal{G}$ for $u_s(t) \in \mathcal{U}$.

The equation

$$i\hbar\dot{V} = \widehat{H}V \tag{1}$$

with the Hamiltonian

$$\widehat{H} = \widehat{H}^0 - \sum_{\mu=1}^{m} u_\mu(t)\widehat{X}^\mu \tag{2}$$

defines on $\mathcal{G}$ a vector field that is right-invariant with respect to the group transformations $\widehat{V} \in \mathcal{G}$. Therefore the system under consideration is right-invariant in the terminology of the publications cited above. If we also take into account the fact that the sphere and therefore the group $\mathcal{G}$ of its unitary transformations into itself are each connected in the natural topology of the Hilbert space of states, we can apply to the given system, for instance, the result proved in Jurdjevič and Sussman (1972) for a broader class of systems and reformulated as follows:

Theorem 1. *If a subgroup $S \subset \mathcal{G}$ associated with a Lie subalgebra $\mathcal{L} \subset L(\mathcal{G})$ generated by operators $\widehat{H}^0, \widehat{X}^1, \ldots, \widehat{X}^m$ is compact or if $\widehat{H}^0 \equiv 0$ (the homogeneous case), then S is the set $\mathcal{A}_U$ of all admissible controlled transformations accessible from the identity of the group $\mathcal{G}$ and corresponding to a class $\mathcal{U}$ of bang-bang controls or the broader class of piecewise continuous controls.*

When the conditions of this theorem hold, the subgroup S is called the *subgroup of controlled transformations*. In view of the fact that the sphere in the state space is connected, it is easy to construct a unitary transformation mapping ψ_0 into ψ_T. Because of the unit normalization of the above vectors, their scalar product can be represented as

$$(\psi_T, \; \psi_0) = e^{i\alpha T} \cos \beta T,$$

where α and β are real numbers. Let us denote by ψ_τ the unit vector directed along the projection of the vector ψ_T onto the subspace orthogonal to ψ_0. Then we can write

$$\psi_T = e^{i\alpha T}(\psi_0 \cos \beta T + \psi_\tau \sin \beta T), \qquad (3)$$

thus taking ψ_T to be the result of rotating the vector ψ_0 in the spanning subspace of the vectors ψ_0 and ψ_τ and multiplying of the vector so obtained by a complex phase factor. This is the required transformation $\widehat{V}(T)$, which can be regarded as the Nth power ($\widehat{V}(T) = \widehat{V}^N(T/N)$) of an elementary transformation $\widehat{V}(T/N)$ acting in a neighbourhood of the identity $\widehat{I}$ of the group $\mathcal{G}$. The radius ϵ_N of this neighbourhood tends to zero as $N \to \infty$, which makes it possible to use the local correspondence of Lie groups and algebras for the investigation of controllability in the large.

The subgroup S is compact in many cases of practical importance, for example, when the domain of controllability is a finite-dimensional sphere.

Then the sufficient condition for complete controllability can be verified by comparing the dimension of the subalgebra L with that of the algebra of all Hermitian matrices representing operators in the state space.

Corollary 1. *If the state space of a quantum system is finite-dimensional, then the condition $\mathcal{L} = L(\mathcal{G})$ is a necessary and sufficient condition for controllability in the large, because the subgroup S coincides with the finite-dimensional group $\mathcal{G}$ and therefore is compact.*

The subgroup S is generally not compact in infinite-dimensional state spaces. It can be appropriate then to make use of another statement from Jurdjevič and Sussman (1972) that is reformulated below to fit the systems in question.

Theorem 2. *A subgroup S is a subgroup of controlled transformations corresponding to a class $\mathcal{U}_f$ of measurable and locally bounded controls if there exists a constant control under which the motion of the system is periodic.*

This theorem enables us to deduce the following inference concerning quantum systems (Braginskii, 1970).

Corollary 2. *Suppose that all the eigenvalues of the operator $\widehat{H}^0$ can be represented as $E_k = kE_0$, where the k are integers and E_0 is an energy level. Then S is the subgroup of controlled transformations corresponding to the class $\mathcal{U}_f$.*

Indeed, at $u_\mu(t) \equiv 0$ in this case the system's motion is periodic, the period being $\tau = 2\pi\hbar/|E_0|$, and therefore the condition of Theorem 2 is satisfied.

The example below is a fairly good illustration of the above general propositions covering the controllability of quantum systems.

Example. Control of the electron's spin. It is well known (Blokhintsev, 1981a, 1981b) that the projection of the electron's spin on a given space axis can only assume two values: $\pm 1/2$. It follows that the state vector $\psi = \{\psi_1, \psi_2\}$ is a two-component one and the spin part of the Hamiltonian in the electric field B with the components B_x, B_y, B_z is a linear combination

$$\widehat{H} = \mu_B(B_x\hat{\sigma}_x + B_y\hat{\sigma}_y + B_z\hat{\sigma}_z)$$

of the 2×2 Pauli spin matrices

$$\hat{\sigma}_x = \begin{pmatrix} 0 & 1 \\ 1 & 0 \end{pmatrix}, \ \hat{\sigma}_y = \begin{pmatrix} 0 & -i \\ i & 0 \end{pmatrix}, \ \hat{\sigma}_z = \begin{pmatrix} 1 & 0 \\ 0 & -1 \end{pmatrix} \tag{4}$$

with coefficient μ_B (the *Bohr magneton*). The three spin matrices form a complete basis of the Lie algebra $ASU(2)$ consisting of 2×2 Hermitian matrices with zero trace.

There are three possible situations:

(A) All three components $B_x = B_x(t)$, $B_y = B_y(t)$, $B_z = B_z(t)$ are piecewise-continuous controls (homogeneous case). Therefore according to Theorem 1, the spin state is completely controllable.

(B) One of the field components is absent, for example, $B_y \equiv 0$, while the two other components $B_x = B_x(t)$ and $B_z = B_z(t)$. This is also a homogeneous case, and in view of Theorem 1 the system is completely controllable because the algebra L generated by any two spin matrices coincides with the algebra $ASU(2)$. It is easy to see this by taking the commutator $\left[\hat{\sigma}_z, \hat{\sigma}_x\right]$ of the matrices $\hat{\sigma}_x$ and $\hat{\sigma}_z$ and forming the linear span of the matrices so obtained, which coincides with $ASU(2)$.

(C) $B_y \equiv 0$ as in the preceding case, but $B_z = \text{const} \neq 0$ (inhomogeneous case), and $B_x = B_x(t)$ is the only control. In view of Theorem 1 and its Corollary 1, the system is controllable in the large because the same Lie algebra $ASU(2)$ arises and its Lie group $SU(2)$ is compact. The same result follows from Theorem 2 since the energy levels of the unperturbed system at $B_x = 0$ have the form $E_m = mE_0$, where $m = \pm 1$.

(D) Only one component of the field is non-zero, for example, $B_z = B_z(t)$, while $B_x \equiv B_y \equiv 0$. This is the case when control is only possible within a one-parameter subgroup of transformations of the form

$$\widehat{V}(t) = \begin{pmatrix} e^{i\phi(t)} & 0 \\ 0 & e^{-i\phi(t)} \end{pmatrix}, \quad \phi(t) = \frac{\mu_B}{\hbar} \int_0^t B_z(\tau)d\tau, \tag{5}$$

which is associated with the Lie algebra $L = \{\alpha\hat{\sigma}_z\}$ with only one parameter α.

In conclusion we note that the solution of the Heisenberg equation

$$i\hbar\frac{d\widehat{A}}{dt} = [\widehat{A}, \widehat{H}^0] - \sum_{\mu=1}^{m} u_\mu(t)[\widehat{A}, \widehat{X}^\mu], \tag{6}$$

defining the evolution of the operator $\widehat{A}$ of a physical variable in the process of control can be presented in the form $\widehat{A}(t) = \widehat{V}(t)\widehat{A}(0)\widehat{V}^{-1}(t)$. Therefore the problem of the accessibility of some final value of the operator $\widehat{A}(T) = \widehat{A}_T$ can be solved by finding the transformation operator $\widehat{V}(T)$ and verifying that it belongs to the subgroup of controlled transformations.

If the unperturbed Hamiltonian is an integral function $\widehat{H}^0 = \widehat{H}^0(\widehat{X}, \widehat{P})$ of the coordinates $\widehat{X}$ and momenta $\widehat{P}$, equations (6) for these operators can be written like the classical Hamilton equations, but in terms of the Heisenberg matrix phase variables. For the controlled Hamiltonian (2), they have the following form:

$$\frac{d}{dt}\widehat{P}^\nu = -\frac{\partial \widehat{H}^0}{\partial \widehat{X}^\nu} + \hat{I}u_\nu(t),$$

$$\frac{d}{dt}\widehat{X}^\nu = \frac{\partial \widehat{H}^0}{\partial \widehat{P}^\nu}.$$

Inasmuch as the control here is carried out not on a sphere but in a matrix-valued phase space, it is possible to apply the well-developed theory of controllability in linear spaces.

It is well known (Kostrikin, 1978) that the rotation group of the space $\mathbf{R}^3$ can be holomorphously embedded into the group $SU(2)$ that has been mentioned in connection with the electron spin state control. In view of (3) and the indicated homomorphism, a change in the electron spin can be imagined as a change in the orientation of the unit vector in $\mathbf{R}^3$ due to a transformation from the group $\mathcal{O}(3)$ and subsequent phase change due to the factor $e^{i\alpha T}$. The transformation from $\mathcal{O}(3)$ can be depicted as in Figure 3.1.1 as a superposition of two rotations $\widehat{\Omega}_1$ and $\widehat{\Omega}_2$ about two different axes $\mathcal{O}_1$ and $\mathcal{O}_2$. It is clear from the figure that whatever the initial point on the sphere might be, the point can be transferred by the phase flows generated by $\widehat{\Omega}_1$ and $\widehat{\Omega}_2$ to any other point of the sphere. This demonstrates complete controllability on the group $\mathcal{O}_3$ provided that the commutator of the rotations $\widehat{\Omega}_1$ and $\widehat{\Omega}_2$ is non-zero: $[\widehat{\Omega}_1, \widehat{\Omega}_2] \neq 0$. The change of phase due to the factor $e^{i\alpha T}$ (the so-called *gradient transformation*, see Rumer and Fet (1970)) does not influence the average value of the observable, in this case the magnetic moment.

3.2. Control on the Motion Groups of Quantum Systems

Recently developed algebraic methods allow us to establish the controllability conditions for dynamical systems that are bilinear with respect to variable states and controls. To this end, one considers the adjoint Lie algebra of the

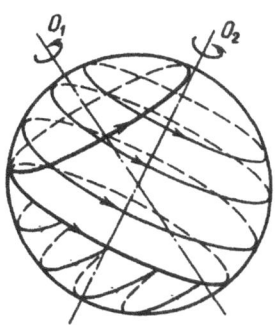

Fig. 3.1.1

right-invariant vector field on a complex group manifold and analyzes the connection between the adjoint Lie algebra and the Lie group of the controlled motion. This approach can also be applied to the investigation of controllability of quantum systems described by an evolution equation, as was shown in the preceding section and earlier in Brockett (1973a). The same approach is considered to be promising in the development of digital automata on controlled transitions of quantum systems (Samoilenko, 1972; Samoilenko and Khorozov, 1980).

The construction of sufficiently efficient algorithms appears to require concrete models of dynamical systems, quality criteria, boundary conditions, as well as the use of transformations reducing the initial problems to forms that are convenient for their solution. In particular, when the controllability group is soluble, there is the possibility of successive optimization based on the decomposition of the group transformations. Sometimes, canonical substitution of dynamical variables makes it possible to go over from group manifolds to linear spaces of states, for which many well-known methods can be applied. We shall follow Samoilenko (1982b) in the presentation of these problems.

Suppose that a system is defined by the Schrödinger equation

$$i\frac{d}{dt}\widehat{U}(t) = \widehat{H}(t)\widehat{U}(t), \ \widehat{U}(0) = I, \tag{1}$$

where the Hamiltonian is

$$\widehat{H}(t) = \widehat{H}_0 + \sum_{\mu=1}^{m}\widehat{H}_\mu u_\mu(t), \tag{2}$$

$\widehat{H}_0, \widehat{H}_1, \ldots, \widehat{H}_m \in \mathcal{K} \subset \mathcal{H}$ being Hermitian matrices of dimension $n \times n$. The minimal Lie algebra $L(\mathcal{H})$ containing these matrices is obviously finite-dimensional. The real functions $u_\mu(t)$, $\mu = 1,\ldots,m$, form a vector-valued control function $\bar{u}(t)\colon R^1 \to \Omega \subset \mathbf{R}^m$. Let us consider the set of all piece-wise smooth vector functions on the closed interval $[0,T]$ to be the class of admissible controls $\mathcal{U}(\Omega)$. As is shown in Brockett (1972), it is possible in many cases to restrict the considerations to the narrower class of staircase controls or even the class of bang-bang controls.

The Euclidean metric in the linear space $\mathcal{K} \subset \mathcal{H}$ of Hermitian matrices $(n \times n)$ is defined by a scalar product of the following form:

$$\langle \widehat{A}, \widehat{B} \rangle = \mathrm{Sp}(\widehat{A}\widehat{B}) = \sum_{k=1}^{n}\sum_{l=1}^{n} A_{kl}B_{kl}^*, \tag{3}$$

where $\mathrm{Sp}(\cdot)$ is the trace of the matrix. Obviously, (3) is real-valued for Hermitian matrices. In the interaction representation (Davydov, 1973; Zubarev, 1971), the evolution matrix can be written as

$$\widehat{U}(t) = e^{-i\widehat{H}_0 t}\widehat{V}(t), \tag{4}$$

where the unitary matrix $\widehat{V}(t)$ is the solution of the differential equation

$$i\frac{d}{dt}\widehat{V}(t) = \Big[\sum_{\mu=1}^{m} e^{i\widehat{H}_0 t}\widehat{H}_\mu e^{-i\widehat{H}_0 t}u_\mu(t)\Big]\widehat{V}(t) \tag{5}$$

with the initial condition $\widehat{V}(0) = \widehat{I}$ ($\widehat{I}$ is the identity matrix) or as the solution of the integral equation

$$\widehat{V}(t) = \widehat{I} - i\int_0^t \Big[\sum_{\mu=1}^{m} e^{i\widehat{H}_0 t_1}\widehat{H}_\mu e^{-i\widehat{H}_0 t_1}u_\mu(t_1)\Big]\widehat{V}(t_1)dt_1, \tag{6}$$

which is equivalent to (5). If the matrix $\widehat{U}(T)$ is given at the right end of the trajectory, then in keeping with (4) we have $\widehat{V}(T) = \exp(-i\widehat{H}_0 T)\widehat{U}(T)$. The problem of finding $\hat{u}(t) \in \mathcal{U}(\Omega)$ given by $\widehat{U}(T)$ or $\widehat{V}(T)$ will be called the *finite control problem* for equations (1) or (5) respectively. Extra conditions, generally that some functional be an extremum, are needed to select a discrete set of solutions of the boundary value problem, in particular, a unique solution. This will be discussed in detail in Chapter 4.

3.3. The Structure of the Algebra of a Quantum System

The accessible sets of bilinear systems were studied in their abstract formulation in Brockett (1972, 1973a, 1973b), in Brockett and Willsky (1975) and in Jurdjevič and Sussman (1972). Below we shall discuss concrete structures with respect to the quantum system (2.1) with the matrix Hamiltonian (2.2) when $\bar{u}(t) \in \mathcal{U}(\mathbf{R}^m)$. On the strength of (2.4), it is sufficient to find out which $\widehat{V}(T)$ can be obtained from (2.5) under every admissible control.

Let us proceed from the operator identity (Brockett, 1973a; Davydov, 1973)

$$e^{\widehat{A}} \widehat{B} e^{-\widehat{A}} = \sum_{k=0}^{\infty} \frac{1}{k!} \operatorname{ad}^k \widehat{A}(\widehat{B}), \tag{1}$$

where

$$\operatorname{ad}^k \widehat{A}(\widehat{B}) = \underbrace{[\widehat{A}, [\widehat{A}, \dots, [\widehat{A}, \widehat{B}] \dots]]}_{k} \quad (\operatorname{ad}^0 \widehat{A}(\widehat{B}) = \widehat{B})$$

is the adjoint representation of the kth order for the element $\widehat{B}$ with respect to the element $\widehat{A}$ in the operator Lie algebra containing $\widehat{A}$ and $\widehat{B}$. In view of (1), equation (2.6) can be expressed as follows:

$$\widehat{V}(t) = \widehat{I} - i \int_0^T \left[\sum_{\mu=1}^m \sum_{s_\mu=0}^{S_\mu} \frac{1}{S_\mu} \operatorname{ad}^{s_\mu} i \widehat{H}_0(\widehat{H}_\mu) t_1^{s_\mu} u_\mu(t_1) \right] \widehat{V}(t_1) dt_1, \tag{2}$$

where the S_μ are the maximum orders of the non-zero adjoint representations. With due regard for the above remark on the finite dimension of the algebra $L(\mathcal{H})$, these finite numbers S_μ exist. The condition $S_\mu \leq n^2 - 1$ must hold in the case of the algebra of $n \times n$ matrices in question.

The basis of the minimum algebra $\Lambda(\mathcal{H})$ containing all the self-adjoint commutators $\operatorname{ad}^{s_\mu} i \widehat{H}_0(\widehat{H})_\mu$ ($s_\mu = 0, 1, \dots, S_\mu$; $\mu = 1, \dots, m$) can be constructed as follows. Given the scalar product (2.3), we can construct the metric real-valued symmetric bilinear form

$$\langle \widehat{H}_\mu, \widehat{H}_\nu \rangle_{s_\mu l_\nu} = \operatorname{Sp}[\operatorname{ad}^{s_\mu} i \widehat{H}_0(\widehat{H}_\mu) \operatorname{ad}^{l_\nu} i \widehat{H}_0(\widehat{H}_\nu)]$$

$$(s_\mu = 0, 1, \dots, S_\mu; \; l_\nu = 0, 1, \dots, S_\nu; \; \mu, \nu = 1, \dots, m). \tag{3}$$

Let its rank be $r_0 \leq m + \sum_{\mu=1}^m S_\mu$. Therefore there exist exactly r_0 linearly independent (in the indicated metric) matrices $\hat{h}_1, \dots, \hat{h}_{r_0}$, that are linear combinations of the commutators $\operatorname{ad}^{s_\mu} i \widehat{H}_0(\widehat{H}_\mu)$. We orthonormalize them by means of any known technique while using (3), and we obtain a set of matrix basis vectors $\widehat{\chi}_1, \ldots, \widehat{\chi}_{r_0}$, which we shall call the initial set and denote by Ξ_0; we denote its linear span by (Ξ_0).

We form from the matrices $\widehat{\chi}_\alpha \in \Xi_0$ the first order commutators

$$\blacksquare \quad \hat{h}_{r_0+1} = \left[\widehat{\chi}_{r_0}, \widehat{\chi}_1\right], \quad \hat{h}_{r_0+2} = \left[\widehat{\chi}_{r_0-1}, \widehat{\chi}_1\right], \ldots$$

$$\ldots, \hat{h}_{2r_0-2} = \left[\widehat{\chi}_3, \widehat{\chi}_2\right], \quad \hat{h}_{2r_0-1} = \left[\widehat{\chi}_2 \widehat{\chi}_1\right],$$

$$\hat{h}_{2r_0} = \left[\widehat{\chi}_{r_0}, \widehat{\chi}_2\right], \quad \hat{h}_{2r_0+1} = \left[\widehat{\chi}_{r_0-1}, \widehat{\chi}_2\right], \ldots,$$

$$\ldots, \hat{h}_{3r_0-3} = \left[\widehat{\chi}_3, \widehat{\chi}_2\right],$$

$$\cdot \quad \cdot \quad \cdot \quad \cdot \quad \cdot \quad \cdot \quad \cdot \quad \cdot \quad \cdot \quad \cdot \quad \cdot$$

$$\hat{h}_{(r_0/2)(r_0+1)} = \left[\widehat{\chi}_{r_0}, \widehat{\chi}_{r_0-1}\right]. \qquad\qquad \square \quad (4)$$

Inasmuch as in the general case (Ξ_0) is not a Lie algebra, the commutant $[(\Xi_0)]$ is not an ideal in (Ξ_0). Therefore the newly constructed operators may prove to be linearly dependent both on each other and on the unit vectors of the initial set.

Assume that the rank of the metric form of the extended set of matrices equals $r_0 + r_1$. Then r_1 of the orthonormalized non-zero matrices of the form

$$\chi_{r_0+\alpha} = \frac{\hat{h}_{r_0+\alpha} - \sum_{\beta=1}^{r_0+\alpha-1} \langle \hat{h}_{r_0+\alpha}, \widehat{\chi}_\beta \rangle}{\|\hat{h}_{r_0+\alpha} - \sum_{\beta=1}^{r_0+\alpha-1} \langle \hat{h}_{r_0+\alpha_0}, \widehat{\chi}_\beta \rangle\|}, \quad \alpha = 1, \ldots, r_1, \qquad (5)$$

supplement the set of matrices $\widehat{\chi}_i$ constructed above. Now let us again construct the commutators of the new unit vectors both with each other and with each unit vector of the preceding set $\hat{h}_\gamma = [\widehat{\chi}_\beta, \widehat{\chi}_\alpha]$, where $\alpha, \beta = 1, \ldots, r_0 + r_1$, $\alpha < \beta$, $\gamma = r_1 + 1, \ldots, \frac{1}{2}r_1(r_1 + 2r_0 + 1)$. We then orthonormalize them by a formula similar to (5).

This process of constructing the commutators and their orthonormalization is continued until a set of commutators is obtained such that each of them is either zero or linearly dependent on the earlier constructed unit (basis) vectors. For a finite-dimensional algebra $L(\mathcal{H})$ and therefore the finite-dimensional adjoint algebra $\Lambda(\mathcal{H})$, the indicated procedure always ends at a certain finite step m_L. Otherwise, the sum $L = r_0 + r_1 + \ldots + r_{m_L-1}$ equal to the dimension of $\Lambda(\mathcal{H})$ would be infinite. For an algebra of $n \times n$ matrices, L is always less than or equal to n^2. The structure constants of the algebra $\Lambda(\mathcal{H})$ with the basis $\widehat{\chi}_1, \ldots, \widehat{\chi}_L$ are given by the numbers

$$c_{\alpha\beta}^{\gamma} = \langle [\widehat{\chi}_\alpha, \widehat{\chi}_\beta], \, \widehat{\chi}_\gamma \rangle. \tag{6}$$

Using these structure constants, it is easy to construct the orthonormalized basis for the commutant of the algebra $\Lambda(\mathcal{H})$. It corresponds to the doubly indexed set of matrices

$$\widehat{\chi}_{\alpha\beta} = \frac{[\widehat{\chi}_\alpha, \widehat{\chi}_\beta]}{\|[\widehat{\chi}_\alpha, \widehat{\chi}_\beta]\|} = \frac{\sum_{\gamma=1}^{L} c_{\alpha\beta}^{\gamma} \widehat{\chi}_\gamma}{\|\sum_{\gamma=1}^{L} c_{\alpha\beta}^{\gamma} \widehat{\chi}_\gamma\|} \quad (\alpha, \beta = 1, \ldots, L; \ \alpha < \beta).$$

The commutants of higher orders can be constructed in the same way. As is well known (Dixmier, 1974), a number of commutants for a finite-dimensional algebra $\Lambda(\mathcal{H})$ form a sequence of ideals embedded into each other

$$\Lambda(\mathcal{H}) = \Lambda^{(0)}(\mathcal{H}) \supset \Lambda^{(1)}(\mathcal{H}) \supset \ldots \supset \Lambda^{(r-1)}(\mathcal{H}) \supset \Lambda^{(r)}(\mathcal{H}). \tag{7}$$

(Recall that a subalgebra J is an *ideal* of an algebra Λ if for any $u \in J$, $v \in \Lambda$ the condition $[u, v] \in J$ holds as well.)

The solution of control problems is essentially simpler in the following cases:

(A) The sequence of commutants terminates in a zero ideal $\Lambda^{(r)}(\mathcal{H}) = (\widehat{0})$ (r is the degree of solubility), that is, the algebra is soluble. (The opposite case is that of a simple algebra when $\Lambda^{(r)}(\mathcal{H}) = \Lambda(\mathcal{H})$.)

(B) The algebra $\Lambda(\mathcal{H})$ can be expanded into a direct sum of prime ideals, in other words, it is semisimple. A criterion for this can be the possibility of expanding the basis $\Xi = \{\widehat{\chi}_1, \ldots, \widehat{\chi}_L\}$ into pairwise commuting disjoint subsets Ξ_q that generate the prime ideals

$$\Xi = \bigcup_{q=1}^{Q} \Xi_q, \ \Xi_p \bigcap \Xi_q \neq \emptyset, [\Xi_p, \Xi_q] = 0, \tag{8}$$

when $p \neq q$ and $[\Xi_p, \Xi_1] \subset (\Xi_p)$, $p, q = 1, \ldots, Q$.

(C) Among the soluble ideals of $\Lambda(\mathcal{H})$, there is a radical R which is a maximal soluble proper ideal containing any soluble ideal $J \subset \Lambda(\mathcal{H})$. Then, as is proved in Dixmier (1974) and in Ovsyannikov (1978), the algebra $\Lambda(\mathcal{H})/R$ is semisimple.

The criteria of solubility and semisimplicity can be sufficiently well-formulated (Dixmier, 1974; Malkin and Manko, 1979) with the aid of the Killing form which is constructed by means of the structure constants (6):

$$k\langle \hat{x}, \hat{y} \rangle = \sum_{\alpha=1}^{L} \sum_{\beta=1}^{L} k_{\alpha\beta} x_\alpha y_\beta, \quad k_{\alpha\beta} = \sum_{\rho=1}^{L} \sum_{\sigma=1}^{L} c_{\sigma\alpha}^{\rho} c_{\rho\beta}^{\sigma}, \tag{9}$$

$$x_\alpha = \mathrm{Sp}(\hat{x}\widehat{\chi}_\alpha), \quad y_\beta = \mathrm{Sp}(\hat{y}\widehat{\chi}_\beta), \quad \alpha, \beta = 1, \ldots, L.$$

The criterion of solubility is

$$k\langle \hat{x}, \hat{x} \rangle = 0 \quad \text{for all } \hat{x} \in \Lambda^{(1)}. \tag{10}$$

The criterion of semisimplicity is

$$\det\{k_{\alpha\beta}\} \neq 0. \tag{11}$$

A more detailed analysis of the structure of the algebra $\Lambda(\mathcal{H})$ can be carried out using the methods suggested in Dixmier (1974) and in Dubrovin et al (1979).

3.4. The Accessible Set of Evolution Matrices

Since the rank of (3.3) has been assumed to be equal to r_0, there are basis sets (one or several) consisting of r_0 linearly independent operators of the form $\mathrm{ad}^s \widehat{H}_0(\widehat{H}_\mu)$. Let us choose a set $r : \{\mathrm{ad}^{s\beta} \widehat{H}_0(\widehat{H}_\mu)$, having the maximum norm

$$\|r\| = \left[\sum_{\beta=1}^{r_0} \langle \mathrm{ad}^{s\mu} i\widehat{H}_0(\widehat{H}_{\mu\beta}), \mathrm{ad}^{s\mu} i\widehat{H}_0(\widehat{H}_{\mu\beta}) \rangle \right]^{1/2}.$$

Then this enables us to realize the whole domain of accessibility for equation (3.2) by the least number of controls equal to r_0 and the minimum mean-square intensity of control. Let us expand the operators of the basis set r in terms of the initial orthonormalized basis $\Xi_0 : \{\widehat{\chi}_1, \ldots, \widehat{\chi}_{r_0}\}$ and then substitute them into (3.2), setting the other matrices equal to zero. It is obvious that the domain of accessibility will not change because the set r has been assumed to be complete. The result is that for a fixed $t = \tau$ we obtain

$$\widehat{V}(\tau) = \hat{I} - i \sum_{\alpha=1}^{r_0} \widehat{\chi}_\alpha \int_0^\tau v_\alpha(t)\widehat{V}(t)dt, \tag{1}$$

where

$$v_\alpha(t) = \sum_{\beta=1}^{r_0} h_{\alpha\beta} t^{s_\beta} u_{\mu\beta}(t), \tag{2}$$

$$h_{\alpha\beta} = \Big\langle \frac{1}{s_\beta!} \mathrm{ad}^{s_\beta} i \widehat{H}_0(\widehat{H}_{\mu\beta}) \widehat{\chi}_\alpha \Big\rangle, \quad \alpha, \beta = 1, \dots, r. \tag{3}$$

Note that because the basis r is complete and the matrices in question are self-adjoint, the matrix $\{\hat{h}_{\alpha\beta}\}$ is real-valued and non-singular.

Now let us suppose that $v_\alpha(t)$ are bounded smooth functions of time and the r_0-dimensional vector of real-valued parameters $\{\xi_1, \dots, \xi_{r_0}\} = \bar{\xi} \in \mathbf{R}^{r_0}$:

$$v_\alpha(t) = v_\alpha(t, \bar{\xi}) = v_\alpha(t, \xi_1, \dots, \xi_{r_0}),$$

as $\{t, \bar{\xi}\}$ varies in some neighbourhood of the origin in the space $\mathbf{R}^1 \times \mathbf{R}^{r_0}$. Then it follows from the theorem on the dependence of solutions on the parameters for the equations of the form (1), that the function $\widehat{V}(\tau) = \widehat{V}(\tau, \bar{\xi})$ is smooth in this neighbourhood. Then by the implicit function theorem, there is a number $\tau > 0$ such that the equation

$$\widehat{W}(\tau, \bar{\xi}) = \hat{I} - i \sum_{\alpha=1}^{r_0} \widehat{\chi}_\alpha \sum_{\beta=1}^{r_0} \xi_\beta \int_0^\tau \frac{\partial}{\partial \xi_\beta} v_\alpha(t, \bar{\xi})\big|_{\xi=0} dt, \tag{4}$$

linearized with respect to $\bar{\xi}$, is solvable in $\bar{\xi}$ for $\widehat{W}(\tau, \bar{\xi}) = \widehat{V}(\tau)$, and then the initial equation (1) is also solvable in $\bar{\xi}$ with the same boundary condition $\widehat{V}(\tau) = \widehat{V}_\tau$.

If the matrix with the entries

$$g_{\alpha\beta} = \int_0^\tau \frac{\partial}{\partial \xi_\beta} v_\alpha(t, \bar{\xi})\big|_{\bar{\xi}=0} dt \tag{5}$$

is non-singular, then we again come to the conclusion that the condition of solvability of (4) is the same as the solvability condition for the problem of finite control described by the equation

$$i\frac{d}{dt} \widehat{X}(t) = \Big(\sum_{\alpha=1}^{r_0} \eta_\alpha \widehat{\chi}_\alpha \Big) \widehat{X}(t), \ \widehat{X}(0) = \hat{I}, \ \widehat{X}(\tau) = \widehat{V}_\tau \tag{6}$$

in terms of the parameters

$$\eta_\alpha = \sum_{\beta=1}^{r_0} g_{\alpha\beta} \xi_\beta.$$

Now let us assume that the controls $u_{\mu\beta}(t) = u_\beta(t, \bar{\xi})$ depend on the vector of the parameters $\bar{\xi}$. In view of (2), the moment equalities (5) take the form

$$g_{\alpha\beta} = \sum_{\gamma=1}^{r_0} L_{\alpha\gamma} \int_0^\tau t^{s\gamma} \frac{\partial}{\partial \xi_\beta} u_\gamma(t, \bar{\xi}) \Big|_{\bar{\xi}=0} dt.$$

Since, by hypothesis, $\det\{h_{\alpha\gamma}\} \neq 0$, the matrix with the entries

$$f_{\alpha\beta} = \int_0^\tau t^{s\alpha} \frac{\partial}{\partial \xi_\beta} u_\alpha(t, \bar{\xi}) \Big|_{\bar{\xi}=0} dt \tag{7}$$

is non-singular, and the matrix $\{g_{\alpha\beta}\}$ turns out to be non-singular as well. However, within the set r of linearly independent operators there cannot be two operators at the same time that have equal indices $s_{\beta_1} = s_{\beta_2}$ and $\mu_{\beta_1} = \mu_{\beta_2}$. Therefore in view of the linear independence of the functions $t^{s\alpha}$ with different s_α, we can find bounded smooth functions $u_\alpha(t, \bar{\xi}) \in \mathcal{U}(\mathbf{R}^m)$ such that the momentum equalities (7) hold for every $\alpha, \beta = 1, \ldots, r_0$ for any given $f_{\alpha\beta}$. In particular, we can demand that $\det\{f_{\alpha\beta}\} \neq 0$.

The following theorem results from these considerations.

Theorem 1. *There exists $\tau = \tau_0 > 0$ such that for $0 \leq t \leq \tau$ any matrix $\widehat{X}(t)$ accessible by virtue of (6) from the identity $\widehat{X}(0) = \widehat{I}$ of the group $G(H)$ is also accessible due to a control $\bar{u}(t) \in \mathcal{U}(\mathbf{R}^m)$ acting upon the system (3.2).*

The construction of the accessibility domain for arbitrary τ can be based on the following statement.

Theorem 2. *Any evolution matrix of the system (2.5) acted upon by the control $\bar{u}(t) \in \mathcal{U}(\mathbf{R}^m)$ can be represented as*

$$\widehat{V}(\theta_1, \ldots, \theta_L) = \exp\left(-i \sum_{\alpha=1}^L \theta_\alpha \widehat{\chi}_\alpha\right), \tag{8}$$

where the θ_α are real parameters and the $\widehat{\chi}_\alpha$ are the matrix basis (unit) vectors of the algebra $\Lambda(\mathcal{H})$.

Proof. Let us introduce into equation (6), instead of the constants $\eta_1, \ldots, \eta_{r_0}$, a staircase control vector $\bar{w}(t) \in \mathcal{U}_c(\mathbf{R}^{r_0})$. According to Theorem 1, for each sufficiently small $(\tau_\nu \leq \tau_0)$ interval I_ν of constant control $\bar{w}(t) = \{w_1^\nu, \ldots, w_{r_0}^\nu\}$,

$t \in \mathcal{T}_\nu$, $\nu = 1, \ldots, N$, the accessible set of the evolution matrices is uniquely determined by the basis set $\Xi_0 = \{\widehat{\chi}_1, \ldots, \widehat{\chi}_{r_0}\}$. By forming all possible products of the form

$$\prod_{\nu=1}^{N} \exp\left\{-i\tau_\nu \sum_{\alpha=1}^{r_0} w_\alpha^\nu \widehat{\chi}_\alpha\right\} \tag{9}$$

from the solutions of (6), we can list every variant of the N-step control. For sufficiently large $N \geq N_0$, the products (9) define a Lie group $G(\mathcal{H})$ whose Lie algebra is $\Lambda(\mathcal{H})$. In view of the well-known correspondence between Lie groups and Lie algebras (Dixmier, 1974; Ovsyannikov, 1978), an arbitrary element of the constructed group can be represented in the form (8). On the other hand, according to Theorem 1, each w_α^ν is associated with a bounded smooth control $u_{\mu\alpha}^\nu(t)$ defined on the interval $\mathcal{T}_\nu$. At the same time, the r_0-dimensional control vector for N steps is bounded and piecewise-smooth, that is $\bar{u}(t) \in \mathcal{U}(\mathbf{R}^{r_0}) \subset \mathcal{U}(\mathbf{R}^m)$. Therefore (8) defines the accessible set of the evolution matrices for every admissible control. This completes the proof.

Note that in order to realize a given matrix (8) it is possible, in principle, to expand $\widehat{\chi}_\alpha$ in terms of the initial matrices $\widehat{H}_0, \widehat{H}_\mu$, and then associate the commutators in the algebra $\Lambda(\mathcal{H})$ with the corresponding commutators in the group $G(\mathcal{H})$. The latter commutators correspond to the control $u_{\mu\alpha}^\nu(t)$, which can be found by solving the moment equalities (7) after the introduction of the parameters $\xi_1, \ldots, \xi_{r_0}$ into the control. The ambiguity in choosing a control, if any, can be avoided by optimization according to the techniques discussed in Chapter 4.

3.5. Designing Discrete Automata on Controlled Transitions of Quantum Systems

Some of the promising trends in the development of computer technology suggest using controlled ensemble processes of a quantum nature (Nikityuk, 1967; Venikov, 1976). The demands for faster performance, better reliability, and increased integration of cybernetic units make it imperative to synthesize automata using quantum processes (Samoilenko, 1972).

It is quite feasible to design quantum discrete automata (QDA), that is, devices whose operation is based on the use of controlled automorphic transformations of discrete sets of quantum system states (Samoilenko and Khorozov,

1980). With this aim in view, the state space of a quantum dynamic system with controlled transitions is subdivided into non-intersecting cells containing vectors of their basis states $|\psi_s\rangle \in \mathcal{S}$ corresponding to the vertices of the given graph $\mathcal{T}$ of the automaton. The set of controls acting upon the system and causing quantum transitions between the basis states defines the input alphabet $\mathfrak{A}$ while the set of macroscopic states of the data output subsystem defines the output alphabet $\mathfrak{B}$.

The functioning of the QDA is determined by two principal operations:

(a) the operation of transition between states $\alpha : \mathcal{S} \times \mathfrak{A} \to \mathcal{S}$ and

(b) the operation of data output $\beta : \mathcal{S} \times \mathfrak{A} \to \mathfrak{B}$.

The QDA can be formally defined by five symbols $\mathcal{H} = (\mathfrak{A},\ \mathcal{S},\ \mathfrak{B},\ \alpha,\ \beta)$. The operations of data input, data output and state control should be regarded as the interaction between micro- and macroscopic processes. Separate subsystems in complex QDA's can also interact directly at the quantum level.

In this approach, since we are dealing with elementary quantum-mechanical processes, an essential role is played not only by the statistical phenomena characteristic for probabilistic automata in general but also by the specific quantum-mechanical effects of superposition of coherent wave functions. The phase relations between the complex amplitudes of state probabilities due to the Heisenberg uncertainty relations must be taken into account as well. Therefore the problem of the physical realization of a discrete automaton by quantum transitions and the feasibility of the deterministic or almost deterministic operations merit special consideration. The crucial role in QDA controllability and observability is played by the operator structure of the models describing the state evolution, whereas the concrete physical nature of the working material used for constructing the QDA and its spatial configuration are inessential for the solution of the problem in principle.

The objective of controlling QDA quantum processes can be reduced to the realization of a prescribed automorphism α of the discrete subset $\mathcal{S}$ of the quantum state space and the realization of the desired function β of the data output from the micro- to the macro-level. The control is accomplished by means of macroscopic influence on the Hamiltonian of the microsystem. These requirements can be accommodated, in particular, by the Hamiltonian of the microsystem

$$\widehat{H} = \widehat{H}_0 + \sum_{l=1}^{L} u_l^\alpha(t)\widehat{X}^l + \widehat{V}(t, q^b(t)), \tag{1}$$

where $\widehat{H}_0$ is the unperturbed Hamiltonian, $u_1^a(t)$ are the input controls (macro-fields), $\widehat{X}^l$ are the microsystem coordinate operators, and $\widehat{V}(t, q^b(t))$ is the potential of the perturbations acting on the microsystem and generated by the macrosystem with the coordinates $q^b(t)$ at the stage of data output. The functions u_l^a and $q^b(t)$ should be regarded here as the letters of the alphabets $\mathfrak{A}$ and $\mathfrak{B}$ respectively. The interaction potential at the stage of transforming the QDA's internal state is assumed to be 'turned off': $\widehat{V}(t, q^b(t)) = 0$, $t_a < t < t_1$. Then the transformation of a state ψ_{s0} into another state ψ_{s_1}, $s_0, s_1 \in S$, defined on the indicated interval, can be represented in the form

$$\psi_{s_1} = \widehat{U}_a(t_1, t_0)\psi_{sa}, \tag{2}$$

where $\widehat{U}_a(t_1, t_0)$ is the evolution operator of the quantum system satisfying the Schrödinger equation

$$i\hbar\frac{\partial\widehat{U}_a(t_1, t_0)}{\partial t} = \left[\widehat{H}_0 + \sum_{l=1}^{L} u_l^a(t)\widehat{X}^l\right]\widehat{U}_a(t_1, t_0) \tag{3}$$

and the initial condition $\widehat{U}_a(t, t_0)\big|_{t=t_0} = \widehat{I}$ ($\widehat{I}$ is the identity operator).

If the quantum system is controllable in the large, the functions $u_l^a(t)$ can be chosen so as to map a finite set S into itself and the system then becomes a discrete automaton. The problem of the completeness of these transformations is closely related to the problem of controllability of dynamic systems on Lie groups.

It follows from Jurdjevič and Sussman (1972) that if the subgroup $S \subset G$ associated with the Lie subalgebra $L \subset \mathcal{L}(G)$ generated by the operators $\widehat{H}_0, \widehat{X}^1, \ldots, \widehat{X}^m$ is compact or if $\widehat{H}_0^0 \equiv 0$, then S is the set of all unitary transformations corresponding to the controls $u_\mu(t)$ from the class $\mathcal{U}$ of piecewise continuous or bang-bang controls.

These conditions are valid for a broad class of physically realizable systems. In particular for spin systems, G is the group $SU(n)$, where n is the dimension of the controllability subspace. For this group, it is possible to indicate the set $\mathcal{D}$ of all discrete subgroups $\mathcal{G}_d \in \mathcal{D}$ and list the sets $\mathcal{U}_d$ of controls that realize each of these subgroups. The subgroup elements $g_d^l \in \mathcal{G}_d$ correspond to quantum transitions between the classes of unitarily equivalent states. Each control vector $u_d^l(t) \in \mathcal{U}_d$ realizes a certain permutation of the initial states. Some of these states are specified by a finite set of values of the observables, which simplifies the coding.

If a QDA is controllable, we can use the structural properties of the automata defined on the groups, construct the desired transition graph $\mathcal{T}$, and so carry out the formal synthesis of the QDA in terms of the given transformation function α.

In order to describe the evolution of the coordinates $q^b(t)$ of the output subsystem, we must use the classical equations

$$\dot{p}^b = -\frac{\partial H^b}{\partial q^b}, \ \dot{q}^b = \frac{\partial H^b}{\partial p^b} \tag{4}$$

with Hamiltonian

$$H^b(p^b, q^b) = H_0^b(p^b, q^b) + \langle \psi_a | \widehat{V}(t, q^b(t)) | \psi_a \rangle, \tag{5}$$

where the interaction potential is averaged by means of the wave function $|\psi_a\rangle$ of the microsystem's state. The subsystems of the QDA's micro- and macro-levels (1)–(3) and (4), (5) prove to be interrelated if $\widehat{V}(t, q^b(t)) \neq 0$. Because the macrosystem must be sensitive to the changes in the microsystem's states, the macrosystem must be in a metastable state, which is defined by the form of the unperturbed Hamiltonian H_0^b.

A reduction of the undesirable influence of thermal fluctuations can be achieved by cooling the QDA to temperatures close to absolute zero. The relaxation processes are thought to be useful not only for the restoration of the automaton's initial state but for the stabilization of the basis states under conditions of parametric excitation.

The cycle of the automaton's performance can be regarded as consisting of the following stages:

(1) relaxation to the starting (ground) state of the QDA when the output subsystem is turned off and control is absent;

(2) a sequence of controls acting upon the automaton in order to transfer it into the given initial state (initial data input);

(3) transformation of states in compliance with the given transition graph by means of a sequence of input controls; and

(4) data transfer from the micro- to macro-level (the output subsystem is turned on and the final state is read out).

At the last stage, the most essential problem is that of the influence of the macrosystem on the microstate in the process of measurement. This process has been described in detail by the model above and can be optimized using the criterion of the minimum probability of an error. The minimum time of reading Δt and the difference between the energy levels ΔE of the output device comply with the Heisenberg uncertainty relation.

Introducing discrete time intervals $t_m = m\Delta t$ and taking into account the group properties of the evolution operator $\widehat{U}(t, t_1)$, we obtain for the multistage transformation $a_n, \ldots, a_1$ the following:

$$\widehat{U}_{a_n \ldots a_1}(t_n, t_1) = \widehat{U}_{a_n}(t_n, t_{n-1}) \ldots \widehat{U}_{a_m}(t_m, t_{m-1}) \ldots \widehat{U}_{a_1}(t_1, t_0).$$

By virtue of the unitarity of $\widehat{U}(t, t_1)$, the control occurs on the sphere of unit radius because of the probabilistic normalization $\langle \psi(t) | \psi(t) \rangle = 1$. As was noted in the preceding sections of this chapter, the processes of controlling a quantum system can be studied on the Lie group G of transformation of the unit sphere of states into itself.

It is important to note that in realizing the desired transition graph it is not enough to bring about a transition from just one state into another. What must be accomplished is to transfer the entire set of states into a preassigned set of states by one and the same control. This is the principal feature of control on groups.

If the spectrum is infinite, controllability in the large does not always take place. However even in this case, there can exist a discrete controllability subgroup of finite dimension N, which is sufficient for the realization of an arbitrary finite graph $\mathcal{T}$ (this is determined by the properties of the operators in the Hamiltonian (1)). By coding the controls by the letters of the input alphabet $a \in \mathfrak{A}$ and establishing a correspondence between the states of the microsystem and those of the macrosystem, we obtain a realization of the transition function α and the output function β. The determination of the finite subgroup of controllability is generally realizable in the resonance approximation.

The consideration of pure states $|\psi\rangle$ is not the only possibility for the description of the performance of a QDA. The control of mixed states given by the density operator $\hat{\rho}$ can also be described by the controlled evolution operator $\widehat{U}$. In this case it is possible to take into account the relaxation processes and estimate the reliability of the performance of the QDA.

A more complicated automaton can be synthesized from separate QDA's if the control operators $\widehat{H}_{ij}(t)$ of the interaction between the individual subsystems are used. By 'turning on' the interaction between the subsystems A_i and A_j and 'turning off' their interaction with the other subsystems, we can first transform the state of this individual pair. Now taking another pair A_j and A_k and considering the state of the subsystems A_j and A_k achieved at the preceding stage to be initial for this stage, we can transform the state of this pair independently from the other subsystems, etc. According to this principle, we can take three subsystems, rather than a pair, and study more complicated unions of them. The transformation of the states of subsystem groups that do not interact with each other can be realized in parallel, without any further complication of the controls.

CHAPTER 4

Optimal Control of Quantum-Mechanical Processes

4.1. General Formulation of the Control Problem for a Quantum Statistical Ensemble

Let us first discuss, following Samoilenko (1982a), the problems of optimal control of a quantum statistical ensemble whose evolution can be defined by the Liouville quantum equation

$$\dot{\rho} = -i[H, \rho] \tag{1}$$

with the initial condition

$$\rho(0) = \rho_0. \tag{2}$$

Here ρ is the statistical operator, H is the Hamiltonian, $[H, \rho]$ is the commutator, and $i = \sqrt{-1}$. Control is accomplished by the direct action of external fields on the system's energy. The technique of thermal control (Kubo, 1957) requires an essentially different approach.

We shall consider only the case of a Hamiltonian that is linear with respect to control fields u_μ

$$H = H_0 + \sum_{\mu=1}^{m} H_\mu u_\mu, \tag{3}$$

where H_0 is the unperturbed Hamiltonian, the H_μ are perturbation operators energetically associated with the fields $u_\mu = u_\mu(t)$, $\mu = 1, \ldots, m$; the latter define a control vector function $\bar{u}(t)$ that belongs to a given class of admissible controls (the details will be given below).

Among the possible functionals of control quality, the most interesting for physical applications are:

(a) the statistical average of the physical variable A, that is, an *observable*, which at time T is equal to the trace Sp of the operator product

$$\overline{A(T)} = \text{Sp}\, A\rho(T); \tag{4}$$

(b) the variance of the observable A at the instant T

$$\overline{\Delta^2 A(T)} = \text{Sp}\, |A|^2 \rho(T) - |\text{Sp}\, A\rho(T)|^2; \tag{5}$$

(c) the average of the observable in the control interval $[0, T]$

$$\bar{A} = \frac{1}{T} \int_0^T \text{Sp}\, A\rho(t)dt; \tag{6}$$

(d) the energy consumption for the control

$$\overline{\|u\|} = \frac{1}{T} \int_0^T \sum_{\mu=1}^m u_\mu^2(t)dt; \tag{7}$$

(e) the time T of achieving the desired final state $\rho(T) = \rho_T$.

When the problem of controllability can be solved, it makes sense to consider optimization problems applied to quantum-mechanical processes and systems. As is known, the theory of optimal control is adequately developed only for linear systems. But the system (1) with Hamiltonian (3) is bilinear with respect to the operator variable ρ and the control $\bar{u}$. Therefore it is necessary to develop new approaches that take this feature into account. For instance, they can be based on the Pontryagin maximum principle and the group analysis of bilinear systems. One such approach is considered below.

4.2. Variational Control Problems

Let us formulate the specific problems of optimal control of a quantum statistical ensemble. First we need to define the mathematical objects involved. We introduce the following notation:

$\mathbf{R}^1$ is the set of all real numbers;

$\mathbf{R}^m = \underbrace{\mathbf{R}^1 \times \ldots \times \mathbf{R}^1}_{m}$ is the set of all possible real number sequences of the length m;

$\mathcal{U}_c(\Omega, T)$ is the set of piecewise smooth control functions $\bar{u}(t) : [0, T] \subset \mathbf{R}^1 \to \Omega \subset \mathbf{R}^m$ with a finite number of points of non-smoothness;

$\mathcal{U}_d(\Omega, T) \subset \mathcal{U}_c(\Omega, T)$ is the class of staircase controls;

$\mathcal{H}$ is a fixed Hilbert space;

$|e_j\rangle$ $(j = 1, 2, \ldots)$ is a fixed basis in $\mathcal{H}$;

$\langle e_k|$ $(k = 1, 2, \ldots)$ is the dual basis of the above;

l_2 is the Hilbert space of square summable sequences of complex numbers;

$F^1(\mathcal{H})$ is the Banach space of all nuclear self-adjoint operators, that is, bounded operators X with finite trace norm

$$\|X_1\| = \mathrm{Sp}|\sqrt{X^*X}| = \sum_j \langle e_j|\sqrt{X^*X}|e_j\rangle < \infty; \tag{1}$$

$\rho(\mathcal{H}) \subset F^1(\mathcal{H})$ is the set of all density operators (statistical operators) in $\mathcal{H}$, that is, the positive self-adjoint operators $\rho > 0$ such that $\mathrm{Sp}\ \rho = 1$; clearly, $\rho(\mathcal{H})$ is a convex set;

$\mathcal{B}(\mathcal{H})$ is the space of all bounded self-adjoint operators in $\mathcal{H}$;

$\mathcal{L}^2(\rho)$ is the Hilbert space of physical quantity operators that are square summable with respect to the state $\rho \in \rho(\mathcal{H})$ (it is also called the *space of observables* associated with the quantum state ρ).

$\mathcal{L}^2(\rho)$ is the completion of $\mathcal{B}(\mathcal{H})$ in the metric given by the scalar product in the following real bilinear form

$$\langle X, Y \rangle_\rho = \mathrm{Sp}\ \rho(X \circ Y^*) = \frac{1}{2}\mathrm{Sp}\ \rho(XY^* + Y^*X) =$$
$$= \mathrm{Re}\ \mathrm{Sp}\ \rho XY^* = \sum_k \rho_k\ \mathrm{Re}\langle X\psi_k|Y^*\psi_k\rangle, \tag{2}$$

where ρ_n and ψ_n are the eigenvalues and eigenfunctions of the operator ρ respectively.

Note that when $Y^* = I$, formula (2) defines for every $\rho \in \rho(\mathcal{H})$ and $X \in \mathcal{L}^2(\rho)$ a real-valued form

$$(X, \rho) = \mathrm{Sp}\ X\rho, \tag{3}$$

that is bilinear in X and convex linear combinations

$$\rho = \theta_1\rho_1 + \ldots + \theta_n\rho_n \quad \forall \theta_n > 0, \ \sum_{k=1}^n \theta_k = 1.$$

Evidently, $X_\rho \in F^1(\mathcal{H})$ and the form (3) can be regarded as the convolution or the scalar product of the indicated operators, the latter being different from the scalar product (2). Therefore the space $\mathcal{L}^2(\rho)$ is conjugate with the set $\rho(\mathcal{H})$ in terms of the convolution (3).

We shall need an essential result established in Kholevo (1980).

Theorem 1 (A.S. Kholevo). *Let $H \in \mathcal{L}^2(\rho_0)$, $\rho_0 \in \rho(\mathcal{H})$, be a time-independent operator. Then $H \in \mathcal{L}^2(\rho_t)$, where*

$$\rho_t = e^{-iHt} \rho_0 e^{iHt} \ \forall t \in (-\infty, \infty). \tag{4}$$

The family $\{\rho_t\}$ is strongly differentiable as a function of time t with values in $F^1(\mathcal{H})$; furthermore, equation (1.1) and the initial condition (1.2) hold.

This statement can be generalized for the case when $H = H_t$ is a staircase operator because individual solutions (4) defined on the intervals where the operator H_t is constant can be coupled in terms of the boundary values.

Corollary. *If H is defined by (1.3), where $\{u_\mu\} = \bar{u}(t) \in \mathcal{U}_d(\Omega, T)$, then on the closed control interval $t \in [0, T]$, the condition $H \in \mathcal{L}^2(\rho_0)$ implies that $H = \mathcal{L}^2(\rho_t)$, where ρ_t is the solution of equation (1.1) under the initial condition (1.2).*

For $u(t) \in \mathcal{U}_c(\Omega, t)$ a similar result can be obtained in the following formulation which is more restricted.

Theorem 2. *If $\bar{u}(t) \in \mathcal{U}_c(\Omega, T)$, $H_0 \in \mathcal{L}^2(\rho_0)$, $\rho_0 \in \rho(\mathcal{H})$, $H_\mu \in \mathcal{B}(\mathcal{H})$, $\mu = 1, \ldots, m$, then $H_0 \in \mathcal{L}^2(\rho_t)$ then $H_0 \in \mathcal{L}^2(\rho_t)$ with respect to the state*

$$\rho_t = U(t) \rho_0 U^{-1}(t), \tag{5}$$

where $U(t)$ is a strongly continuous unitary group which is the solution of the uniformly well-posed Cauchy problem

$$\dot{U} = -iHU, \ U(0) = I \tag{6}$$

for the Schrödinger equation with the identity as the initial operator.

Proof. Since $\mathcal{U}_d(\Omega, T)$ is everywhere dense in $\mathcal{U}_c(\Omega, T)$, it follows that for all $\epsilon_k > 0$, $\bar{u}(t) \in \mathcal{U}_c(\Omega, T)$ there exists $\bar{u}^{(k)} \in \mathcal{U}_d(\Omega, T)$ such that $\|\bar{u}(t) - \bar{u}^{(k)}\| < \epsilon_k$.

Therefore the difference between the Hamiltonians H and $H^{(k)}$ corresponding to the controls $\bar{u}$ and $\bar{u}^{(k)}$ has the estimate

$$\|H - H^{(k)}\| \leq \sum_{\mu=1}^{m} \|H_\mu\| \, |u_\mu - u_\mu^{(k)}| < \epsilon_k \sum_{\mu=1}^{m} \|H_\mu\|.$$

Since $H_\mu \in \mathcal{B}(\mathcal{H}) \; \forall \mu \in [1, m]$, it follows that $H^{(k)} \to H$ as $\epsilon_k \to 0$. Therefore the sequence $U^{(k)}(t)$ of solutions for the uniformly well-posed Cauchy problem (Krein, 1964) with the staircase Hamiltonians $H^{(k)}$ converges uniformly to the strongly continuous group $U(t)$ which is the solution of the Cauchy problem (6). On the interval $[0, T]$, except for the points where the controls are not smooth, the derivative $\dot{U}(t)$ is determined uniquely; therefore in view of (6), differentiation of (5) brings us to (1.1).

Since the domain $\mathcal{D}(H)$ of the operator H in (3) does not depend on t and the theorem holds for all $\bar{u}^{(k)}(t) \in \mathcal{U}_d(\Omega, T)$ as follows from the corollary, the statement of the theorem also holds for the limit vector-valued control function $u(t) \in \mathcal{U}_c(\Omega, T)$. This completes the proof.

Note that in view of (2) it can be stated that if $X \in \mathcal{L}^2(\rho)$, then for any $\rho \in \rho(\mathcal{H})$ scalar products $\langle X, X \rangle$ and $\langle X, I \rangle_\rho$ are defined having the sense of the variance and the expected value of the observable X, because it is obvious that $I \in \mathcal{L}^2(\rho)$ for all $\rho \in \rho(\mathcal{H})$. Consequently, the functional (4) is defined if and only if the functional $\mathrm{Sp}\,|A|^2 \rho(T)$ is defined, where A is the operator of the observable. In particular, if $H_0 \in \mathcal{L}^2(\rho_0)$, the functional (1.4) is meaningful for any $\bar{u}(t) \in \mathcal{U}_d(\Omega, T)$ or $u(t) \in \mathcal{U}_c(\Omega, T)$ (Theorems 1 and 2). In the general case, in order to guarantee the existence of the functionals (1.4)–(1.6), it is necessary to postulate that A belongs to the spaces $\mathcal{L}^2(\rho_t)$ at the appropriate $t \in [0, T]$.

We shall list here the most typical formulations of the optimal control problems for a quantum statistical ensemble.

A. **The problem of Mayer.** Minimize the functionals (1.4) and (1.5) for given T when $\Omega = \Omega_Q$ is the closed convex hull of the vectors $u^q \in \mathbf{R}^m$ ($q = 1, \ldots, Q$), $\bar{u}(t) \in \mathcal{U}_c(\overline{\Omega}_Q, T)$.

B. **The problem of Bolza.** Minimize the functional (1.6) for $T = \mathrm{const}$ and $\bar{u}(t) \in \mathcal{U}_c(\overline{\Omega}_Q, T)$.

C. **The terminal time-optimal problem.** Bring the system from ρ_0 to ρ_T in the minimum time T for $\bar{u}(t) \in \mathcal{U}_c(\overline{\Omega}_Q, \infty)$.

D. **The terminal problem with a quadratic criterion.** Bring the system from ρ_0 to ρ_T for $T = $ const and the minimum value of the functional (1.7), assuming that $\bar{u}(t) \in \mathcal{U}_c(\mathbf{R}^m, T)$. Of interest also are linear combinations of (1.7) with other functionals.

Equation (1.1) in the problems listed above appears as an equation involving a differential. Therefore each of them is a problem of finding a conditional extremum. Instead of (1.1), we can use the expression (2.5) together with equation (2.6). If the conditions of Theorem 2 are satisfied, the above formulations of the problems are well-posed in the sense of continuous dependence of their solutions on the initial data and parameters.

4.3. Necessary Conditions for an Extremum

Let us discuss the core of the problem using problem A (the problem of Mayer). In order to determine the Hamilton function, we prove the following statement.

Lemma. *Let $\rho_0 \in \rho(\mathcal{H})$ and assume that either of the two conditions holds:*

a) $H_\mu \in \mathcal{L}^2(\rho_0) \quad \forall \mu \in [U, m], \quad \bar{u}(t) \in \mathcal{U}_d(\Omega, T)$

or

b) $H_0 \in \mathcal{L}^2(\rho_0), \quad H_\mu \in \mathcal{B}(\mathcal{H}) \quad \forall \mu \in [1, m], \quad \bar{u}(t) \in \mathcal{U}_c(\Omega, T)$.

If $X_t = U(t)X_0 U^{-1}(t)$, where $U(t)$ is the solution of the Cauchy problem (2.6), then the condition $X_0 \in \mathcal{L}^2(\rho_0)$ implies that $X_t \in \mathcal{L}^2(\rho_t)$. Furthermore, the convolution $\mathrm{Sp}\, X_t \rho_t$ is defined for all $t \in [0, T]$ and satisfies the Lagrange identity

$$\mathrm{Sp}\, X_t \rho_t = \mathrm{Sp}\, X_0 \rho_0 = \text{const.} \tag{1}$$

Proof. As follows from Theorem 2 and the Corollary to Theorem 1, if either condition (a) or (b) is valid, the unitary group $U(t)$, which is the solution of the Cauchy problem (2.6) on the interval $[0, T]$, maps the spaces $\mathcal{L}^2(\rho_0)$ and $F^1(\mathcal{H})$ into themselves. If $\rho_0 \in \rho(\mathcal{H})$, $X_0 \in \mathcal{L}^2(\rho_0)$ then $X_0 \rho_0 \in F^1(\mathcal{H})$ and the convolution $\mathrm{Sp} X_0 \rho_0$ is defined. But the trace is invariant with respect to unitary transformations of the space $F^1(\mathcal{H})$ into itself. Therefore $\rho_t = U(t)\rho_0 U^{-1}(t)$ and $X_t = U(t)X_0 U^{-1}(t)$, whence

$$\mathrm{Sp}\, X_0 \rho_0 = \mathrm{Sp}\, U(t)X_0 \rho_0 U^{-1}(t) = \mathrm{Sp}\, U(t)X_0 U^{-1}(t)U(t)\rho_0 U^{-1}(t) = \mathrm{Sp}\, X_t \rho_t,$$

which completes the proof.

We now derive necessary conditions for a minimum of the functional (1.4), in the presence of the differential connection with the initial condition (1.2), in the form of the maximum principle (Pontryagin et al., 1963; Moiseev, 1971; Rozonoer, 1959; Butkovskiy, 1969). Suppose that $\bar{u}^0(t) \in \mathcal{U}_c(\Omega, T)$ is an optimal control. Taking an infinitesimal interval $\tau - \epsilon < t < \tau$, where $\epsilon > 0$, we replace $\bar{u}^0(t)$ by another control $\bar{u}(t) \in \mathcal{U}_c(\Omega, T)$ without any change in the control beyond this interval. At the point τ, the statistical operator ρ has the variation

$$\delta\rho_\tau = -i\epsilon \sum_{\mu=1}^{m} [H_\mu, \rho_\tau](u_\mu(\tau) - u_\mu^0(\tau)). \tag{2}$$

We denote by $X_t \in \mathcal{L}^2(\rho_t)$ the operator associated with the quantum state ρ_t and satisfying both the equation

$$\dot{X}_t = -i[H, X_t] \tag{2}$$

and the boundary condition at the right end of the trajectory $X_t : [0, T] \to \mathcal{L}^2(\rho_t)$:

$$X_T = -A. \tag{4}$$

Here the Lagrange identity holds. Indeed, by virtue of (1.1) and (3)

$$\frac{d}{dt} \mathrm{Sp}\, X_t \rho_t = \mathrm{Sp}\dot{X}_t \rho_t + \mathrm{Sp}\, X_t \dot{\rho}_t = -i\mathrm{Sp}[H, X_t]\rho_t -$$
$$-i\mathrm{Sp}\, X_t[H, \rho_t] = -i\mathrm{Sp}[H, X_t]\rho_t - i\mathrm{Sp}[X_t, H]\rho_t \equiv 0.$$

Equation (3) can be regarded as a consequence of the more general relation

$$X_t = U(t)X_0 U^{-1}(t), \tag{5}$$

that was used in the lemma proved here.

The variation of the functional (1.4) can be expressed in terms of the variation of the right end of the trajectory of $\rho_t : [0, T] \to \rho(\mathcal{H})$. Because $\overline{\delta A(T)}$ must be a real non-negative quantity, we can use (1), (2) and (4) to obtain

$$\overline{\delta A(T)} = \mathrm{Sp}\, A\delta\rho_T = -\mathrm{Sp}\, X_T \delta\rho_T = -\mathrm{Sp}\, X_\tau \delta\rho_\tau =$$
$$= i\epsilon\mathrm{Sp}\, X_\tau \sum_{\mu=1}^{m} [H_\mu, \rho_\tau](u_\mu(\tau) - u_\mu^0(\tau)) \geq 0.$$

Since $\epsilon > 0$, this means that

$$-i\mathrm{Sp}\,X_\tau \sum_{\mu=1}^{m}[H_\mu, \rho_\tau]u_\mu^0(\tau) \geq -i\mathrm{Sp}\,X_\tau \sum_{\mu=1}^{m}[H_\mu, \rho_\tau]u_\mu(\tau). \qquad (6)$$

Let us introduce into our consideration the Hamilton function, which can be obtained as the scalar product of the form (1) in the right hand side of (1.1) and the variable X conjugate to ρ:

$$h = -i\mathrm{Sp}[\rho, X]H = -i\mathrm{Sp}[H, \rho]X.$$

It is real-valued because the operator $-i[\rho, X]$ is self-adjoint and the trace of the product of two self-adjoint operators is real. On the strength of this, h may also be represented as

$$h = \mathrm{Im}\,\mathrm{Sp}\{[\rho, X]H(\bar{u})\}. \qquad (7)$$

In view of (1.3), it is easy to see that the inequality (6), which is valid at any instant $t \in [\epsilon, T]$ for an arbitrarily small $\epsilon > 0$, implies that the following maximum principle holds.

Theorem. *Assume that $\bar{u}^0(t) \in \mathcal{U}_c(\Omega, \tau)$ is an optimal control minimizing the function (1.4) in the presence of the differential connection (1.1) with the initial condition (1.2). Then it is necessary that there exist a function $X_t :$ $[0, T] \to \mathcal{L}^2(\rho_t)$ associated with the state $\rho_t \in \rho(\mathcal{H})$ and satisfying equation (3) with the final condition (4), such that at any instant $t \in [0, T]$ that is a point of continuity of the control $\bar{u}^0(t)$, the function (7) of the variable $\bar{u} \in \Omega$ attains its maximum at the point $\bar{u} = \bar{u}^0(t)$.*

The proof of this theorem follows the general maximum principle for equations in Banach spaces and its essentials are contained in the above derivation with due regard for the remarks on the conjugate spaces and the lemma.

Since the first term in (1.3) does not depend on the control, we have the following corollary.

Corollary. *The optimal control is determined by the formula*

$$\bar{u}^0(t) = \arg\max_{\bar{u} \in \Omega} \sum_{\mu=1}^{m} \mathrm{Im}\,\mathrm{Sp}\{[\rho_t, X_t]H_\mu\}u_\mu, \qquad (8)$$

if and only if $[\rho_0, X_0] \neq 0$. *In particular, for* $\Omega = \overline{\Omega}_Q$ (*see the problem of Mayer above*), $\bar{u}^0(t) = \bar{u}_0^q$, *where*

$$q_0 = q_0(t) = \arg \max_{q \in [1,Q]} \sum_{\mu=1}^{m} \text{Im Sp}\{[\rho_t, X_t]H_\mu\}u_\mu^q. \tag{9}$$

This necessitates choosing one of the Q functions, namely, the one that possesses the greatest value at the given instant.

Therefore, using the maximum principle in the problem of Mayer, we arrive at a boundary problem for equations (1.1) and (3) with the boundary conditions (1.2) and (4), where the control is eliminated by means of (8) or (9). A specific feature of the problems with the Liouville equations is that instead of solving them it is sufficient to solve equation (2.6) and use (2.4) and (2.5). In a similar way we can transform the commutator

$$[\rho_t, X_t] = U(t)[\rho_0, X_0]U^{-1}(t), \tag{10}$$

in order to find $\bar{u}^0(t)$.

The derivation of necessary conditions for an extremum for the other functionals considered above may be carried out in the same manner without any fundamental change except for a few points. For instance, take the problem of optimal performance rate. The boundary condition (4) is replaced by the condition $\rho(T) = \rho_T$. The problem of Bolza can be reduced to the problem of Mayer with the functional $\bar{A} = \text{Sp}\, A\sigma(T)$ and the additional equation $\dot{\sigma} = \rho/T$ under the initial condition $\sigma(0) = 0$. If the combined criterion

$$J = \text{Sp}\, A\rho(T) + \frac{1}{T} \int_0^T \sum_{\mu=1}^{m} u_\mu^2(t)dt$$

is applied in the problem with the free end of the trajectory, the optimal control must be a critical point of the extended Hamilton function. This gives rise to the explicit expression for the components of the vector-valued function $\bar{u}(t)$:

$$u_\mu^0(t) = \frac{T}{2} \text{Im Sp}\{[\rho_t, X_t]H_\mu\}, \quad \mu \in [1, m].$$

The problems for the functionals (1.4) and (1.5) merely involve a different boundary condition for the conjugate variable.

4.4. Methods of Solving Boundary Value Optimization Problems

A practical technique of finding optimal control consists in the solution of a boundary problem which the original variational problem can be reduced to using the maximum principle. As was mentioned above, the state variables and the conjugate variables in a Liouville quantum ensemble are operators. This complicates the system's description and the representation of its states while, at the same time, the integration of the equations becomes more stable because all the eigenvalues of the evolution operators are equal to one in modulus.

For the numerical representation of the operators, it is usually convenient to go over from the space $\mathcal{H}$, where the initial description of the system is given, to the space l_2, where the operators are expressed as matrices. Selecting suitable bases $|e_j\rangle$ and $\langle e_i|$, we obtain the matrix elements of the Hamiltonian $H_{ij} = \langle e_i|H|e_j\rangle$ and similarly, of the other operators. We suppose this substitution of spaces to be carried out. Therefore in what follows the operators are interpreted as matrices and we shall denote them by the same letters.

A convenient method of solving the boundary value problem corresponding to the problem of Mayer is the following version of the method of invariant immersion (Bulatov, 1977) which is based on a gradual lengthening of the interval T of control. According to (3.9), $\bar{u}^0(t) \in \mathcal{U}_d(\Omega_Q, T)$, where Ω_Q consists of given vectors $\bar{u}_q$. These vectors are associated with the evolution operators $U_q(t)$ which are the solutions of the Cauchy problem

$$\dot{U}_q = -i\Big(H_0 + \sum_{\mu=1}^{m} H_\mu u_\mu^q\Big)U_q(t), \quad U_q(0) = I.$$

These operators (matrices) must be calculated in advance, before the boundary value problem is solved.

The intervals of constant control will be denoted by $\tau_1, \ldots, \tau_{N(t)}(t)$, where $N(t)$ is a time-dependent integer, so that

$$t = \sum_{\alpha=1}^{N(t)} \tau_\alpha.$$

For an arbitrary fixed sequence of controls numbered $q_\alpha \in [1, Q]$, $\alpha = 1, \ldots$ $\ldots, N(t)$, we can write

$$U(t, q_\alpha, \tau_\alpha) = U_{q_{N(t)}}(\tau_{N(t)}) \ldots U_{q_1}(\tau_1). \tag{1}$$

In view of the boundary condition (3.4), instead of (3.10) we obtain

$$[p_t, X_t] = U(t, q_\alpha, \tau_\alpha)[p_0, X_0]U^{-1}(t, q_\alpha, \tau_\alpha) =$$
$$= U(t, q_\alpha, \tau_\alpha)[p_0, U^{-1}(T, q_\alpha, \tau_\alpha) \times$$
$$\times AU(T, q_\alpha, \tau_\alpha)]U^{-1}(t, q_\alpha, \tau_\alpha). \qquad (2)$$

The control numbers q_α are determined by (3.9) for each constant control interval τ_α.

Since (2) is continuous in t, for small enough $T \leq T_1$ there exists only one constant control interval ($N(T) = 1$). Under the same condition, one of the functions

$$Z^{q_1}(t) = \sum_{\mu=1}^{m} \operatorname{Im} \operatorname{Sp}\{U(t, q_1)[U^{-1}(T, q_1)AU(T, q_1), p_0]U^{-1}(t, q_1)H_\mu\}u_\mu^{q_1} \qquad (3)$$

maintains its greatest value over the whole interval $t \in [0, T]$. Now $U(T, q_1) \to I$ as $T \to 0$ and therefore

$$Z^{q_1}(t) \to Z^{q_1} = \sum_{\mu=1}^{m} \operatorname{Im} \operatorname{Sp}\{[A, p_0]H_\mu\}u_\mu^{q_1} \qquad (4)$$

uniformly in t. Selecting the greatest number from the Z^{q_1} ($q_1 \in [1, Q]$) we can find the optimal value $q_1 = q_1^0$ (assuming that $[A, p_0] \neq 0$; the case of $[A, p_0] = 0$ corresponds to a special control which is not considered here).

We then proceed as follows. For $q_1 = q_1^0$ we gradually increase T, at the same time comparing the functions $Z^{q_1}(t)$ ($q_1 \in [0, 1]$, $t \in [0, T]$). If T reaches the value given in the problem and q_1^0 stays optimal, then the solution has already been found, since q_1^0 determines the evolution operator and the variables p_t and X_t. Otherwise, starting from some $T = T_1$, the condition that $Z^{q_0}(t)$ be a maximum will fail to hold, while at the beginning it will fail to hold in a small ϵ-neighbourhood of the point $t_1 \in [0, T_1]$. Inside this neighbourhood, (3) will be maximized by a control possessing another number $q = q_2^0 \neq q_1^0$. We then apply the corresponding evolution operator over the indicated interval and repeat this iterative procedure until the desired duration of the control interval $[0, T]$ is achieved.

4.5. Methods of Direct Optimization on Unitary Groups

Other optimization techniques can be based on direct methods of variational calculus and the expansion of the infinitesimal generating group $U(t)$ with respect to some basis. By way of illustration, let us consider the possibility of reversible energy exchange between a controlling field and a system that is initially in a state of thermodynamic equilibrium given by the Gibbs distribution

$$\rho_0 = \exp\left[\frac{1}{\theta}(FI - H_0)\right], \tag{1}$$

where θ is temperature, F is free energy, I is the identity matrix, and H_0 is the non-perturbed Hamiltonian. If the system is controllable on the group $U(n)$ (Brockett, 1972; Butkovskiy and Samoilenko, 1980), its evolution operator can be represented as

$$U(t) = \exp\left(-i\sum_{\alpha=1}^{L} \theta_\alpha(t)\chi_\alpha\right), \tag{2}$$

where L is the dimension of the Lie algebra $AU(n)$ (Rumer and Fet, 1970), $\{\chi_\alpha\}$ is the orthonormal basis and the $\theta_\alpha(t)$ are real-valued functions of time. Then the statistical operator at an instant t can be expressed as

$$\rho_t = e^{A_t}\rho_0 e^{-A_t} = \rho_0 + \frac{1}{1!}[A_t, \rho_0] + \frac{1}{2!}[A_t[A_t, \rho_0]] + \ldots, \tag{3}$$

where

$$A_t = -i\sum_{\alpha=1}^{L} \theta_\alpha(t)\chi_\alpha.$$

Assume that the control beyond the control interval $[0, T]$ equals zero. Then after applying the control field, the average energy of the statistical ensemble can be expressed as

$$\overline{E}_T = \mathrm{Sp}\,(H_0\rho_T). \tag{4}$$

This is the functional (1.4) with $A = H_0$. We denote the initial value of the average energy by $\overline{E}_0 = \mathrm{Sp}\,(H_0\rho_0)$. In view of (3) and (4), the energy increment to within terms of order greater than two in the parameters $\theta_\alpha(T) = \theta_\alpha$ is given by the formula

$$\Delta\overline{E} = \overline{E}_T - \overline{E}_0 = \frac{1}{2}\sum_{\alpha=1}^{L}\sum_{\beta=1}^{L}\theta_\alpha\theta_\beta\mathrm{Sp}\{[H_0, \chi_\alpha][\rho_0, \chi_\beta]\} + \ldots \tag{5}$$

Further analytical consideration requires a concretization of the group $U(n)$. The controllability on the group $SU(2)$ was discussed in §3.1, therefore let us use here the more complicated group $SU(3)$. A basis of the algebra

$ASU(3)$ is $\chi_\alpha = \frac{\sqrt{2}}{2}\lambda_\alpha$, $\alpha = 1,\ldots,8$, where the λ_α are the Gell-Mann matrices employed in elementary particle theory. The whole set of matrices is presented below:

$$\lambda_1 = \begin{pmatrix} 0 & 1 & 0 \\ 1 & 0 & 0 \\ 0 & 0 & 0 \end{pmatrix}, \; \lambda_2 = \begin{pmatrix} 0 & -i & 0 \\ i & 0 & 0 \\ 0 & 0 & 0 \end{pmatrix}, \; \lambda_3 = \begin{pmatrix} 1 & 0 & 0 \\ 0 & -1 & 0 \\ 0 & 0 & 0 \end{pmatrix},$$

$$\lambda_4 = \begin{pmatrix} 0 & 0 & 1 \\ 0 & 0 & 0 \\ 1 & 0 & 0 \end{pmatrix}, \; \lambda_5 = \begin{pmatrix} 0 & 0 & -i \\ 0 & 0 & 0 \\ i & 0 & 0 \end{pmatrix}, \; \lambda_6 = \begin{pmatrix} 0 & 0 & 0 \\ 0 & 0 & 1 \\ 0 & 1 & 0 \end{pmatrix}.$$

$$\lambda_7 = \begin{pmatrix} 0 & 0 & 0 \\ 0 & 0 & -i \\ 0 & i & 0 \end{pmatrix}, \; \lambda_8 = \frac{1}{\sqrt{3}}\begin{pmatrix} 1 & 0 & 0 \\ 0 & 1 & 0 \\ 0 & 0 & -2 \end{pmatrix}.$$

Calculations according to (5) yield the quadratic form of the control parameters θ_α

$$\Delta\overline{E} = -\frac{1}{2}[(\theta_1^2 + \theta_2^2)(E_1 - E_2)(\rho_1 - \rho_2)+$$
$$+(\theta_4^2 + \theta_5^2)(E_1 - E_3)(\rho_1 - \rho_3) + (\theta_6^2 + \theta_7^2)(E_2 - E_3)(\rho_1 - \rho_3)]. \quad (6)$$

Here $E_1 < E_2 < E_3$ are the eigenvalues of the non-perturbed energy operator H_0; ρ_1, ρ_2, ρ_3 are the eigenvalues of the statistical operator (5) which is a function of H_0 and therefore can be reduced to diagonal form in the energy representation along with the energy operator.

The expression (6) permits us to make the following conclusions.

1) The parameters θ_3 and θ_8 have no influence on the result, therefore it is sufficient in order to control $\Delta\overline{E}$ to employ in the quadratic approximation the appropriate 6-parameter subgroup of the group $SU(3)$.

2) If ρ_1, ρ_2, ρ_3 do not form a monotonically non-increasing sequence, that is, if $\rho_1 < \rho_2$, then it is possible to control the removal of energy from the system and the transfer of the energy to the control field. Otherwise, for example, in the distribution (1), this is not feasible.

4.6. Maximization of the Probability of Observing a Given State of a Quantum System

The theory of optimal processing of quantum signals arouses considerable interest in connection with the development of quantum electronics. It is well known (Neumann, 1932) that the measurement of a physical variable gives a reliably unique value if the wave function $|\psi\rangle$ describing the state of the system is an eigenfunction of the operator of the observable $\widehat{A}$. In particular, this occurs when the Hamiltonian $\widehat{H}_1(t)$ of the interaction between the quantum system and the measuring instrument at a certain instant commutes with the operator of the observable quantity $\widehat{A}$ (Helström, 1976; Kholevo, 1980). From the viewpoint of control theory, the problem of bringing the system into the subspace of the indicated states can be regarded as the problem of controlling the spectrum in terms of reducing the Hamiltonian to diagonal form in the representation of the eigenfunctions of the operator of observables or *vice versa*. The condition $[\widehat{A}, \widehat{H}_1] = 0$ is sufficient for non-perturbing measurement of $\widehat{A}$, but this does not eliminate the measurement error on a finite time interval in view of the uncertainty relation.

The approach presented below was suggested in Khorozov (1980). The idea is that the purposeful control of the quantum states of a system can essentially influence the result of measurement. The problem is to find the states that yield the maximum probability that the observable is measured accurately.

In (p, q)-space we consider a quantum system with L degrees of freedom, the system occurring in the state $|\psi\rangle$ described in the coordinate representation by a wave function that is finite in the momentum representation. Assume that the generalized coordinates $\{q_l\}$ are a complete set of pairwise commuting observables and possess a continuous spectrum of eigenvalues.

The expected value of the result of measurement equals the probability that the value a of the observable A is in an interval Δ. Therefore in order to accomplish measurement it is sufficient to represent the observable A in terms of the projection operator

$$\Pi_\Delta = \int_\Delta |a\rangle\langle a| da,$$

for instance,

$$\Pi_R = \int_R |\{q_i\}\rangle\langle\{q_i\}| \prod_m dq_m,$$

where Δ is the set of eigenvalues a of the operator $\widehat{A}$ which can only take the two values, 1 or 0. Measuring Π_Δ, we obtain a reading of 1 or 0 depending on whether or not the value of A is within the interval Δ.

The problem is to maximize the functional

$$\langle\psi|\Pi_R|\psi\rangle = \int_R \prod_{j=1}^{L} dq_j |\langle\{q_l\}|\psi\rangle|^2 \tag{1}$$

under the extra condition

$$\langle\psi|\Pi_D|\psi\rangle = \int_D \prod_{j=1}^{L} d\rho_j |\langle\{p_l\}|\psi\rangle|^2 = 1, \tag{2}$$

where $\langle\{p_l\}|\psi\rangle$ is the representation function of the same state in terms of the continuum of the basis vectors $|\{p_l\}\rangle$ of the momentum representation of the observables:

$$R = \{q : (2n_i - 1)\xi\frac{a}{2} \le q_i \le (2n_i + 1)\xi\frac{a}{2}\},$$

$$D = \{p : (2m_i - 1)\frac{\hbar}{4a} \le p_i \le (2m_i + 1)\frac{\hbar}{4a}\}.$$

The parameter ξ characterizes the dilatation of a phase space cell. The ranges of integration are selected such that the volume of the phase space cell is equal to $(\hbar/2)^{2L}$. The condition (2) follows from the assumption that $[\widehat{A}, \widehat{H}_1] = 0$.

A similar mathematical problem is known in optics and radiophysics (Razmakhin and Yakovlev, 1971). With the aid of the earlier developed techniques, we obtain the integral equation

$$\mu(\xi)\langle\{p_L\}|\psi\rangle \prod_{j=1}^{L} e^{ip_j n_j \xi(a/\hbar)} =$$

$$= \int_D \prod_{j=1}^{L} dp_j \frac{\sin(p_j - p'_j)\xi(a/2\hbar)}{\pi(p_j - p'_j)}\langle\{p'_L\}|\psi\rangle \prod_{j=1}^{L} e^{-ip'_j n_j \xi(a/\hbar)}. \tag{3}$$

The function that maximizes the expected value can be found through the eigenfunction of the integral equation (3) and represented in spheroidal functions. It is easy to show that

$$\max[\langle\psi|\Pi|\psi\rangle] = \mu_0(\xi) = \frac{\xi}{4\pi}[R_{0,0}^{(1)}(\xi, 1)]^2.$$

The asymptotic expression for $\mu_0(\xi)$ in the one-dimensional case is

$$1 - \mu_0(\xi) = \sqrt{2\pi\xi}e^{-\xi/4}\Big[1 - \frac{3}{4\xi} + O\Big(\frac{1}{\xi^2}\Big)\Big].$$

It is clear from equation (3) that any phase space cell gives a wave function that differs from the wave function of the central cell by a phase factor. The system of functions so generated can be taken as the basis. The coefficients of expansion of an arbitrary function in terms of this basis yield the probability amplitudes that the system occurs within a cell. Therefore if it is assumed that the system is controllable, the control process transfers the space of the system's states to the space of the obtained states, the expansion coefficient becoming equal to unity. Since we used projection operators in the 'measurement', we can speak of it as almost deterministic measurement.

Depending on the purpose of control, one can pose the problem of finding the states of the quantum system that either minimize the uncertainty (the minimum variance) or maximize the probability that the results of measurement are accurate. In the former case, we can employ coherent states (utilizing the sign of the equality in the uncertainty relation); in the latter case we can employ the states mentioned above. Therefore the preference of either case depends on the technique of estimating the measurement error.

CHAPTER 5

Dynamical Systems with Stored Energy and Negative Susceptibility

5.1. The Effect of Negative Susceptibility of Dynamical Systems and its Applications

Dynamical systems, when their internal degrees of freedom are excited, can occur in a state of negative static susceptibility to external forces. This state is characterized by the fact that the generalized displacement, being locally time-averaged, proves to be directed oppositely with respect to the static or slowly varying generalized force causing the displacement. This chapter discusses possible applications of this phenomenon in order to stabilize unstable objects; conditions will be obtained under which dynamical systems with stored energy, occurring in either a state of stable equilibrium or finite motion, possess negative generalized static susceptibility. Such systems can be used to establish fast negative feedback with amplification (Samoilenko, 1978a). This chapter also discusses examples of both classical and quantum systems of various physical natures that feature such properties.

The *generalized susceptibility* of a dynamical system (Davydov, 1973) is the characteristic of the system's response to an external influence. This response can be mathematically expressed as the dependence $x = \hat{\alpha}f$ of the vector x of generalized coordinates on the vector f of generalized forces, where $\hat{\alpha}$ is the generalized susceptibility operator. If the system's inertia is inessential for its description, the notion of *static susceptibility* can be used. Generally, static susceptibility is defined by a tensor α_{ij} whose principal values α_k are real if the generalized coordinates and forces are real. This makes it possible to talk

117

about the *sign of static susceptibility in specified directions of the coordinate space*. If this sign is constant in every direction or if the direction is fixed, we talk about *global susceptibility*.

When the external conditions specify coordinates rather than forces, it is convenient to use the reciprocal of the generalized susceptibility. We term it 'generalized resistivity' and define it by the relation $f = \hat{c}x$, where $\hat{c}$ is the operator of generalized resistivity and $\hat{c} = \hat{\alpha}^{-1}$.

We shall need the following conditions of physical realizability of the susceptibility (or resistivity) operator:

a) the induced component of the output variable does not appear earlier than the input variable;

b) the operator is uniformly continuously dependent on its parameters;

c) the signs of the increments of the input and output variables are the same at the beginning of the transition period.

The condition c) characterizes those operators in question for which the product of the input and the output variables defines the measure of energy transferred from external bodies or fields to the given system. Therefore c) means that the sign of the initial perturbing energy, actuating the internal mechanism of the system's susceptibility, is positive. In particular, the linear stationary operator of susceptibility or resistivity defined by a transfer function in the form of the rational fraction

$$W(p) = \frac{b_m p^m + \ldots + b_0}{a_n p^n + \ldots + a_0},$$

can be considered to be physically realizable if (a) $n \geq m$, and (b) the numerator and the denominator do not possess the same roots in the right half-plane of the complex variable p, and (c) $b_m/a_n > 0$.

Moreover, stability or at least finiteness of motion is required for the long-term physical existence of a state of the system with a certain operator of susceptibility or resistivity.

Generalized coordinates and forces are regarded as external if they appear in the description of the system's interaction with external objects, and the rest of the coordinates and forces are regarded as internal. The specification of external conditions means either fixing the external coordinates and force vectors or setting up operator relations between them that together with the initial conditions define the system's motion uniquely. The behaviour of a given system in its interaction with external objects broadly depends on the system's autonomous features. It is convenient to consider the fixation of the input variables, that is, the coordinates or forces, as the standard conditions of isolation of the system.

The sign of the system's susceptibility depends on its physical nature. There are examples of natural continuous media that are stable in fixed external fields and possess either positive or negative static susceptibility. It is well known (Landau and Lifshits, 1959) that if the system is in thermodynamic equilibrium, then its electric susceptibility is always positive while its magnetic susceptibility χ can be either positive (in para- or ferromagnetism) or negative (in diamagnetism). The absolute value of susceptibility of natural diamagnetics in thermodynamic equilibrium never exceeds unity. The value of susceptibility $\chi = -1$ can only occur in superconductors and these are sometimes regarded as superdiamagnetics (Vonsovskii, 1971). The term 'superdiagmagnetism' can also be applied to the phenomena of anomalously strong magnetic susceptibility $\chi < -1$, the possibility of which was shown in Samoilenko (1974) and in Samoilenko and Khorozov (1977); this will be discussed in this chapter. A superdiamagnetic state can be of practical value for the implementation of distributed systems of control of a magnetic field (Samoilenko, 1974).

The purpose of the present discussion is to investigate the general conditions for the appearance of negative susceptibility and its various macroscopic manifestations. This can be largely useful for the realization of fast negative feedback with amplification or fast actuators controlling the energy-consuming inertial systems with lumped and/or distributed parameters (Butkovskii, 1965, 1975, 1977, 1979a, 1979b, 1982; Butkovskii and Samoilenko, 1979a, 1979b, 1980; Butkovskii and Pustylnikov, 1980; Butkovskii and Pustylnikova, 1982).

In boundary control of the processes in continuous media (Ladikov-Roev, 1978; Samoilenko, 1968), there occurs a situation in which both control of the system and collection of data on the state of the system can take place in the same space-time domain of contact between the control device and the controlled system. In automatic control systems, the functions of control and collection of data on the controlled system are carried out by different units, actuators and sensors. In order to increase the performance rate and improve the space resolution of the distributed control device (Butkovskii, 1979a), it is advisable to join these functions in the same distributed unit. From the formal viewpoint, it is sufficient to join the variables of the plant (the controlled system) in the boundary domain of its contact with the control device by means of appropriate operator relations. These relations are called *impedance relations* and are determined from the given objective of control. They are realized by means of the control device connected to the plant (Samoilenko, 1968).

Boundary control is a possible but not the only impedance approach that will be considered here. For instance, consider the system

$$A_1 \ddot{x}_1 + B_1 \dot{x}_1 + C_1 x_1 = C x_2,$$
$$A_2 \ddot{x}_2 + B_2 \dot{x}_2 + C_2 x_2 = C^T x_1, \tag{1}$$

where A_i, B_i, C_i are matrices, x_i are vectors, $i = 1$ stands for the plant, $i = 2$ stands for the controller, and the superscript T denotes transposition. The real-valued matrix C of interaction between the plant and the controller can be singular. The potential energy of the interaction can be represented in the bilinear form $x_1^T C x_2 = x_2^T C^T x_1$ and the constraints by the generalized force vectors

$$f_1 = C x_2, \ f_2 = C^T x_1. \tag{2}$$

The *susceptibility* of any of the subsystems, in particular, the controller, is the physical characteristic of the response of the subsystem to the action of another subsystem on it. The *susceptibility operator* (*susceptibility* for short) expresses the generalized displacement x in terms of the generalized force f. For a linear stationary controller, the susceptibility operator $\hat{\alpha}_2$ is the matrix of transfer between the vectors f_2 and x_2 defined by the second equation in (1):

$$x_2 = \hat{\alpha}_2 f_2. \tag{3}$$

The operator that is inverse to the susceptibility operator $(\hat{Z}_2 = \hat{\alpha}_2^{-1})$, is called the *impedance operator*, or the *impedance* for short.

Both susceptibility and impedance play the principal role here in formulating the laws of feedback in the system. Indeed, we can eliminate x_2 in the first equation of (1) with the aid of (3) and (2) to obtain the equation of the closed control system in the following form:

$$A_1 \ddot{x}_1 + B_1 \dot{x}_1 + C_1 x_1 = C \hat{\alpha}_2 C^T x_1.$$

Here the feedback law is defined by the operator $\hat{R} = C \hat{\alpha}_2 C^T$ which depends on the susceptibility operator of the controller; if the rank of the matrix C

equals the dimension of the vector x_1, then the spectrum of the matrix $\widehat{R}$ is contained in the spectrum of the matrix $\hat{\alpha}_2$. Control of the spectrum is treated in Vinogradskaya and Girko (1980) and in Girko and Vinogradskaya (1979).

In synthesizing such systems there is much to be gained by a preliminary analysis of the influence of the mathematical structure of the forces on the stability of motion (Merkin, 1971). For instance, it follows from the first Tait-Thomson theorem (Thomson and Tait, 1879) that gyroscopic stabilization is impossible in a potential system with an odd number of unstable coordinates; if the number of these coordinates is odd, it is better to start with a change in this number and only then take measures to damp oscillations. For this reason, in the classical gyroscopic stabilization of a monorail car, for example, the one described in Merkin (1971), the system is supplemented by an unstable degree of freedom, mainly by a small weight fixed at the gyrocompass outer gymbal in the extreme top position.

An analysis of the force structure in the system indicates another possible technique of stabilization. If the operator $\widehat{R}$ could be synthesized in the form

$$R = U + V\frac{d}{dt},$$

the system's stability would be easy to maintain by an appropriate choice of the matrices U and V with the necesssary number of negative eigenvalues. However, the trouble is that it is not quite obvious that it is physically possible to realize a susceptibility operator with negative eigenvalues such that the controller possessing it would not alter the number of roots of the characteristic equation of the open system in the domain of instability, that is, the controller would be stable itself.

Note that the negative kinetic susceptibility defining the matrix V reveals itself in the well-known phenomena of negative conductivity or negative kinematic viscosity (negative friction) and finds its broad application in generators, amplifiers, and certain mechanical devices. Therefore we shall only consider the realization of negative static susceptibility defining the matrix U. This phenomenon has not come into wide use yet and is scarcely applied other than in the known cases of electrical circuits with a negative capacity (Gertsenshtein et al., 1971), negative inductance (Samoilenko, 1973; Kuchtenko and Samoilenko, 1973; Artemenkov et al., 1971) and the application of the fundamental possibility of achieving the superdiamagnetic state in ferromagnetic crystals by means of longitudinal pumping (Samoilenko, 1974; Samoilenko and Khorozov, 1977). It will be shown below that the state of negative static susceptibility can occur in dynamical systems of various physical natures under certain conditions.

5.2. Synthesis of Bipolar Circuits with Negative Impedance and Negative Conductivity

Given an operational voltage amplifier with sufficiently fast performance, one can synthesize a bipolar circuit (Figure 5.2.1) possessing a negative impedance in the effective frequency range and stable under no-load conditions. An elementary analysis yields

$$U = U_1 - U_2 = -[K(p) - 1]U_1 = -[K(p) - 1]Z(p)I, \tag{1}$$

where U is the voltage on the bipolar circuit, U_1 and U_2 are the input and output of the amplifier, $K(p)$ is its transfer function ($p = d/dt$, or $p = i\omega$ in the frequency analysis), $Z(p)$ is the impedance of the auxiliary signal bipole and I is the current produced by an external source, the current passing through the bipole $Z(p)$ and the output circuit of the amplifier. It follows that $U = Z_e(p)I$, where $Z_e(p) = -[K(p) - 1]Z(p)$ is the equivalent impedance of the bipolar circuit. The sign of this impedance for real $K(p) = k$ turns out to be opposite to the sign of $Z(p)$ if $k > 1$. Therefore, if $Z(p)$ is an ordinary physically realizable impedance, $Z_e(p)$ is also a realizable negative impedance.

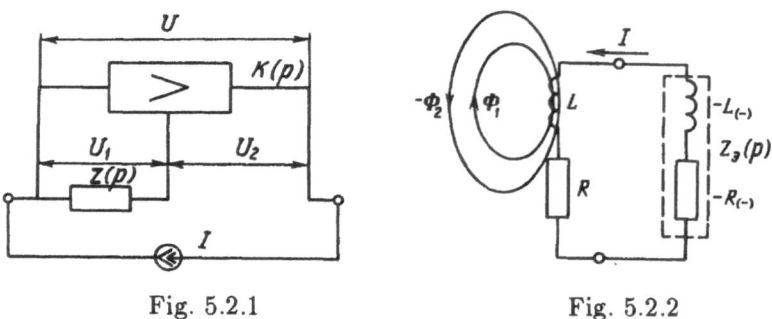

Fig. 5.2.1   Fig. 5.2.2

The poles of the impedance $Z_e(p)$ are the poles of the amplifier and the signal bipole. Therefore, if each of them is stable, then the bipolar circuit is stable under no-load conditions, that is, when the feedback loop through the external circuit is disconnected from the bipolar circuit. If the external circuit does not pass high frequencies, an analysis of stability only requires a low-frequency circuit with the equivalent impedance while the characteristic of the amplifier should be replaced by an ideal charateristic with the real amplification factor k. In particular, if $Z(p) = Lp + R$, then $Z_e(p) = -L_{(-)}p - R_{(-)}$, where $L_{(-)} = (k+1)L$ and $R_{(-)} = (k-1)R$.

If the inductance-resistor circuit is magnetically coupled with the stabilized system (Figure 5.2.2), we can obtain a stable amplifier of the magnetic field with a negative flux reflection factor. The parameters of the equivalent dipole can be chosen so as to compensate completely for the active resistance of the coupled circuit and partially for the inductance by setting $R_{(-)} = R$, $L_{(-)} < L$. An analysis of the circuit (Figure 5.2.2) shows that the ratio of the secondary magnetic flux Φ_2 to the primary one Φ_1 will be negative and its absolute value will be greater than unity:

$$\frac{\Phi_2}{\Phi_1} = \frac{LI}{\Phi_1} = \frac{L}{\Phi_1} \frac{-p\Phi_1}{p(L - L_{(-)}) + R - R_{(-)}} = -\frac{L}{L - L_{(-)}} < -1. \qquad (2)$$

This technique of field amplification found application in the maintenance of equilibrium in a toroidal plasma pinch (Samoilenko, 1973; Artemenkov et al., 1971; Kuchenko and Samoilenko, 1973).

The operational amplifier in the example above had an infinite input impedance and zero output impedance, that is, it was regarded as an ideal voltage amplifier with positive current feedback. The negative voltage feedback, which is not shown explicitly in Figure 5.2.2, can be used in order to decrease non-linear and frequency distortions in the operational amplifier.

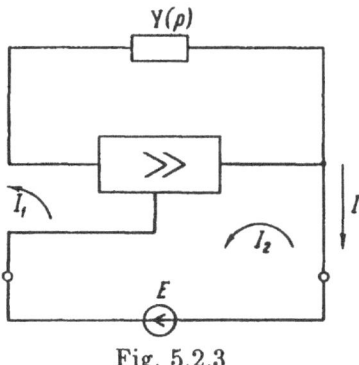

Fig. 5.2.3

Now let us consider an operational current amplifier with a zero input impedance and a positive voltage feedback through a bipole with conductivity

$Y(p)$, as is shown in Figure 5.2.3. This is a bipolar circuit realizing a negative conductance. In the notation of Figure 5.2.3, we can write the following relations which are dual to (1):

$$I = I_1 - I_2 = -[K(p) - 1]I_1 = -[K(p) - 1]Y(p)E.$$

Hence $I = -Y_e(p)E$, where $Y_e(p) = -[K(p) - 1]Y(p)$ is the equivalent conductance of the given bipolar circuit. If the real current amplification factor $K(p) = k > 1$ is large enough, the conductance is negative. In particular, if $Y(p) = Cp$ is a capacitor, we obtain a bipolar circuit with a negative capacitance $C_e = C_{(-)} < 0$, for which $Y_e(p) = -(k - 1)Cp = -C_{(-)}(p)$. This circuit is stable if shorted, provided the amplifier and the auxiliary bipole are both stable.

It is not difficult to synthesize an amplifier of an electric field on the basis of a bipolar circuit with a negative capacitance as is shown, for instance, in Gertsenshtein et al., (1971). For the purpose of automatic control, it is essential that the element of negative capacitance type should have a negative susceptibility in the sense discussed above, the generalized coordinate being the electric charge passing through the bipolar circuit and the generalized force being the voltage applied. This feature makes it possible to use a negative capacitance for automatic stabilization of electrostatically unstable systems.

5.3. Negative Susceptibility in Gyroscopically Related Systems

When there is the problem of controlling fast processes, such as the processes in charged particle beams, the usual operational amplifiers of transfer type with separate input and output may prove to be unsuitable because of the large delay of signals in these amplifiers. The use of such amplifiers in this and other similar cases may also be technically impracticable, since too many parallel channels are required for distributed control. These factors stimulate the search for the realization of negative susceptibility without transfer-type amplifiers, which would make it possible to carry out amplification of energy density with reflection of minimum delay.

In this connection it is of interest to discuss gyroscopically coupled systems defined by heterogeneous equations of the form

$$A\ddot{x} + HG\dot{x} + \Lambda x = f. \tag{1}$$

Here x and f are vectors of the generalized coordinates and forces, A, G and Λ are symmetric, skew-symmetric and diagonal matrices respectively, and H is the gyroscopic parameter.

Let us find out, on the basis of general theorems on the influence of the structure of forces on the stability of motion (Merkin, 1971; Thomson and Tait, 1879), what kind of forces are necessary to accommodate the following to demands at the same time:

1) in the static approximation, at least one susceptibility coefficient must be negative, for example, $\alpha_1 = x_1/f_1 < 0$;

2) the system must be stable in the Lyapunov sense.

Evidently, the first requirement is satisfied if the first element of the matrix Λ is negative: $\lambda_1 = -c^2 < 0$. Then, on the strength of the first Tait-Thomson theorem, there must be other negative elements of the matrix as well, so that the total number of negative eigenvalues would be even and the second requirement would be met. While these conditions are met, the two demands above are accommodated at the same time in the absence of multiple eigenfrequencies and if H is large enough.

A very simple system accommodating the two demands is

$$a^2\ddot{x}_1 - c^2 x_1 - Hg\dot{x}_2 = f_1, \quad Hg\dot{x}_1 + a^2\ddot{x}_2 - c^2 x_2 = 0, \tag{2}$$

which is associated with the dynamic susceptibility

$$\alpha(p) = \frac{a^2 p^2 - c^2}{(a^2 p^2 - c^2)^2 + H^2 g^2 p^2} \tag{3}$$

and the operator of the gyroscopic coupling between x_2 and f_1 in the form $x_2 = -\gamma(p)f_1$, where

$$\gamma(p) = \frac{Hgp}{(a^2 p^2 - c^2) + H^2 g^2 p^2}. \tag{4}$$

The static susceptibility is negative: $\alpha(p)|_{p=0} = -\alpha_{(-)} = -1/c^2 < 0$ and at the same time, the system is stable in the Lyapunov sense for $Hg > 2ac$, because $\alpha(p)$ possesses first-order purely imaginary poles:

$$p_\pm^\pm = \pm \frac{iHg}{\sqrt{2}a^2}\sqrt{1 - \frac{2a^2 c^2}{H^2 g^2} \pm \sqrt{1 - \frac{4a^2 c^2}{H^2 g^2}}}.$$

Let us couple the synthesized system with the controlled unstable potential system

$$a_0^2 \ddot{x}_0 - c_0^2 x_0 = -k x_1,$$

$$H g \dot{x}_2 + a^2 \ddot{x}_1 - c^2 x_1 = -k x_0, \qquad (5)$$

$$-H g \dot{x}_1 + a^2 \ddot{x}_2 - c^2 x_2 = 0.$$

Here x_0 is the generalized coordinate of the controlled system, c_0 is its negative resistivity and k is the constant of potential connection between the two systems (a relation via x_1 only). The matrix of the potential forces

$$\Pi = \begin{bmatrix} -c_0^2 & k & 0 \\ k & -c^2 & 0 \\ 0 & 0 & -c^2 \end{bmatrix}$$

possesses three (that is, an odd number) negative coefficients of resistivity in its diagonal. However, this does not mean that the number of negative eigenvalues cannot be even, which is necessary for stability according to the first Tait-Thomson theorem. Indeed, if $k^2 > c^2 c_0^2$, that is, if $k^2 \alpha_{(-)} > c_0^2$, the matrix Π possesses only two negative eigenvalues.

There is no doubt now that, given the gyroscopic parameter H is great enough, the system will be stable in the Lyapunov sense. This simple example allows us easily to obtain this result directly from the characteristic determinant of the system (5), which yields the conditions of neutral stability in the form $Hg > 2ac$, $k > cc_0$. But we have obtained these results by demanding, firstly, the stability of the control system and, secondly, a large enough value of the negative static susceptibility $\alpha_{(-)} > c_0/k^2$ of this system. From this last inequality we can draw the physically obvious inference: when the coupling factor k decreases or the negative resistivity c_0 increases, the absolute value of the negative susceptibility $\alpha_{(-)}$ of the stabilizing system increases. A decrease in $\alpha_{(-)}$ and k brings the system back to instability.

If the coupled control system is neutrally stable, perturbations will accumulate in it and sooner or later depart from the linear mode of operation, which may bring about the loss of its stabilizing properties. It is difficult to make it asymptotically stable by physically realizable means. Therefore, the system is only suitable for the stabilization of relatively short-lived processes in an immobile medium or for the stabilization of processes in a moving medium of small dimensions, for example, an accelerated ensemble of charged particles.

In the latter case, a large number of stabilizing systems should be placed along the trajectory of the ensemble.

As a very simple stabilizer, we can use any system possessing low inertia, a mechanical moment of momentum, and an electric or magnetic moment. The magnetic moment provides for a convenient electromagnetic coupling with the stabilized system.

5.4. Transverse Susceptibility of a Rigid Dipole in an Inversely Directed Constant Field

One possible technique (although not the only one, see Merkin, 1974) of introducing generalized gyroscopic forces into a system is the use of elements possessing an angular momentum, that is, appropriate gyroscopic elements. By supplementing such an element by an electric (or magnetic) moment, we can obtain a gyroelectric (or gyromagnetic) element.

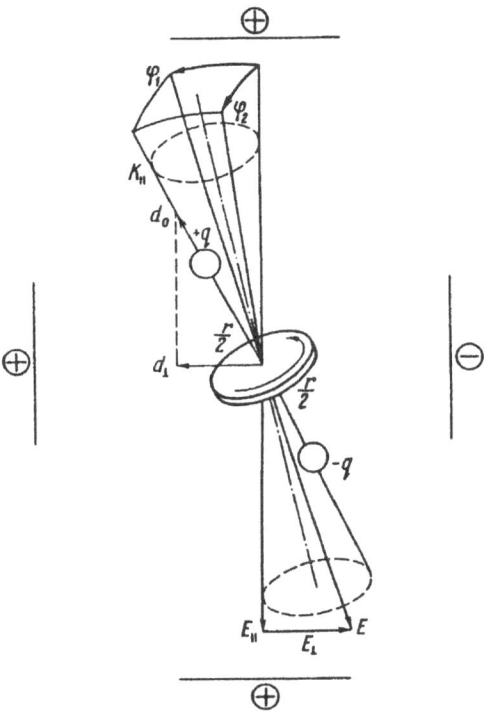

Fig. 5.4.1

For the sake of definiteness let us discuss a gyroelectric element and consider its realization as a means of realizing negative electric susceptibility (Figure 5.4.1). Two charges $+q$ and $-q$ are placed at the ends of a rigid rod of the length l. A flywheel rotates around the rod. The vector $K_\parallel$ of the flywheel's angular momentum has the absolute value $K_\parallel = J_\parallel \omega_\parallel$, where $J_\parallel$ is the axial moment of inertia and $\omega_\parallel$ is the angular velocity of the flywheel. This electromechanical dipole, possessing an electric moment d_0, whose absolute value is $d_0 = ql$, is placed between the plates of two capacitors put at right angles to each other. One of them gives a constant longitudinal electric field $E_\parallel$ in the reverse direction to the non-perturbed dipole moment, while the other creates a field $E_\perp$ that is transverse to the non-perturbed dipole moment. The system can also have a third capacitor creating the component E_2 of the electric field at right angles to $E_\parallel$ and $E_\perp$ (Figure 5.4.1 does not show this capacitor). The analysis is first carried out for only one transverse component $E_\perp$; the result will then be given for the general case.

Owing to the presence of the opposite (reversely directed) longitudinal field $E_\parallel$, the rigid dipole in the non-perturbed state possesses a positive potential energy $U_0 = E_\parallel d_0$, which the dipole is able to give to external systems and then restore depending on the direction of the transverse field. The longitudinal moment of momentum K imparts stability to the dipole in the inversely directed field. Let us consider this in detail and find the linear susceptibility of the given gyroelectric element.

For small angular deviations ϕ_1 and ϕ_2 of the dipole axis from the direction of the longitudinal field, the differential equations of motion are

$$\dot\phi_0 = \omega_\parallel = \text{const},$$
$$J_\parallel \omega_\parallel \dot\phi_2 + J_\perp \ddot\phi_1 - E_\parallel d_0 \phi_1 = -E_1 d_0, \qquad (1)$$
$$-J_\parallel \omega_\parallel \dot\phi_1 + J_\perp \ddot\phi_2 - E_\parallel d_0 \phi_2 = 0,$$

where $J_\perp$ is the transverse moment of inertia of the electromechanical dipole. The projection of the electric moment onto the transverse plane is the system's transverse polarization. The components of polarization for small angles of deviation are

$$P_1 = \phi_1 d_0, \quad P_2 = \phi_2 d_0. \qquad (2)$$

Introducing these new variables, we can obtain from the last two equations of

(1) the following:

$$Hg\dot{P}_2 + a^2\ddot{P}_1 - c^2 P_1 = E_1, \quad -Hg\dot{P}_1 + a^2\ddot{P}_2 - c^2 P_2 = 0, \tag{3}$$

where

$$Hg = J_\parallel \omega_\parallel / d_0^2, \quad a^2 = J_\perp / d_0^2, \quad c^2 = E_\parallel / d_0. \tag{4}$$

Obviously this system is the same as (3.2) if the transverse components of the polarization vector are assumed to be the generalized coordinates and the transverse electric field E_1 is the generalized force. The above correspondence between the dynamical variables enables us, using (3.3) directly, to make a general inference about the negative value of the static susceptibility $\alpha(p)|_{p=0} = -\alpha_{(-)}$, where $\alpha_{(-)} = d_0/E_\parallel$ and obtain the condition of stability $\omega_c < \frac{1}{2}\sqrt{U_0/J_\perp}$ where $\omega_c = c^2/Hg = U_0 = E_\parallel d_0$ is the precession frequency. If there is also a second transverse component E_2, the susceptibility can be expressed by the tensor

$$\hat{\alpha}(p) = \begin{bmatrix} \alpha(p), & -\gamma(p) \\ \gamma(p), & \alpha(p) \end{bmatrix},$$

where $\alpha(p)$ and $\gamma(p)$ are defined according to (3.3) and (3.4).

The fact that the transverse electric moment that is in the reverse direction to the transverse field can be explained qualitatively as follows. The gyroscope in a force field is known to precess with respect to the field direction. If the field varies adiabatically, the axis of precession follows the direction of the field (Figure 5.4.1). In the non-perturbed motion, the dipole moment, which is electric in this case, may have a direction either coinciding with the field direction or opposite to it. We are choosing the latter. Now if the inclination of the precession axis is caused by a transverse field E_1, the vector of the dipole moment will precess around a new axis while possessing a pulsating component $d_\perp$ which, on the average, is in the reverse direction to $E_\perp$.

The energy of interaction of the dipole with the longitudinal field, while the precession axis is inclined, decreases on the average over the period of precession by the value of $\Delta\overline{U} = U_0(\bar{\phi}_1)^2$. The energy of interaction with the transverse field is negative, because the susceptibility is negative, and on the average equals $\overline{U}_1 = \overline{P}_1 E_1$. It follows from (3) and (4) that

$$E_1 = -c^2 \overline{P}_1 = \frac{-E_\parallel}{d_0} \overline{P}_1.$$

In view of (2), we therefore obtain

$$\overline{U}_\perp = -\frac{E_\parallel}{d_0}(\overline{P}_1)^2 = -E_\parallel d_0(\bar{\phi}_1)^2.$$

This corresponds in value to $\Delta\overline{U}_0$. Thus, there is an automatic transfer of energy from the longitudinal field to the transverse one when the transverse field is intensified, but this process reverses when the transverse field diminishes. In both cases the energy is in the reverse direction to the external perturbations.

We shall not analyze here the relaxation processes destroying the state of negative susceptibility of a dipole in an inversely directed field. The case of inverse magnetization of a gyromagnetic medium is treated in Samoilenko (1974). It is also shown there that there is the possibility of sustaining long-term negative susceptibility in such a medium by using the technique of parametric pumping. In Samoilenko and Khorozov (1977), pumping with a complicated harmonic composition is suggested in order to prevent instability of the spin waves. Both in these publications and in the example above, negative susceptibility is obtained using natural gyroscopic forces. Below we discuss another possibility, where generalized gyroscopic forces are produced as a result of parametric excitation of a system that does not have its own angular momentum in the non-perturbed state.

5.5. Negative Susceptibility of a Parametrically Modulated Oscillator

Suppose that the rigid electric dipole considered in the preceding section does not have its own angular momentum created by a rotating flywheel. Now let the constant electric field E_0 be oriented upwards, so that the dipole in the equilibrium state possesses the minimum potential energy and is an oscillator that can have angular oscillations in a plane (Figure 5.5.1). The constant longitudinal field is superimposed with a small periodic field $E_\sim$ possessing the first and second harmonics of the natural frequency of the non-perturbed oscillator. The field $E_\parallel = E_0 + E_\sim$ retains its longitudinal orientation. The field $E_\perp$ acts in the transverse direction.

Taking damping forces into account, the equation of dipole oscillations in dimensionless coordinates can be written as follows:

$$\frac{d^2y}{dt^2} + 2\epsilon^2\eta\frac{dy}{dt} + (1 + 2\epsilon^2\xi)y + \epsilon y\cos t + \frac{\epsilon^2}{2}y\cos 2t = x_0, \tag{1}$$

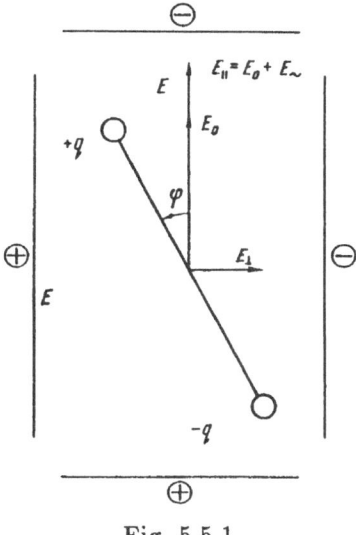

Fig. 5.5.1

where y is the transverse component of the dipole's electric moment (the transverse polarization), x_0 is the transverse electric field, η is the damping, ξ is the deviation of the oscillator's frequency from the pumping principal frequency, $0 < \epsilon \ll 1$ is a small parameter characterizing the amplitude of pumping and t is time.

It is convenient to look for the solution of (1) by the method of harmonic balance (the balance of slowly varying amplitudes) in the form

$$y = y_0 + \epsilon^{-1}(u_1 \cos t + v_1 \sin t) + u_2 \cos 2t + v_2 \sin 2t + \epsilon u_3 \cos 3t + \epsilon v_3 \sin 3t + O(\epsilon^2).$$
(2)

In substituting (2) into (1), we assume that the harmonic amplitudes of forced oscillations vary in time at the rates of the order of ϵ^2, as does the external field x_0. The balance in harmonics and in the parameter ϵ yield the following system of curtate equations

$$y_0 = x_0 + \frac{1}{4}u_1, \quad (Hp + C + P)W = X,$$
(3)

where

$$W = \begin{bmatrix} u_1 \\ v_1 \end{bmatrix}, X = \begin{bmatrix} X_0 \\ 0 \end{bmatrix} \text{ and } X_0 = -\frac{x_0}{2}$$

are the vectors of the generalized coordinates and forces,

$$H = \begin{bmatrix} 0 & 1 \\ -1 & 0 \end{bmatrix}, \quad C = \begin{bmatrix} -\gamma & 0 \\ 0 & -\gamma \end{bmatrix}, \quad P = \begin{bmatrix} 0 & \eta \\ -\eta & 0 \end{bmatrix}$$

are the matrices of the gyroscopic, potential, and non-conservative position-sensitive forces respectively, and

$$\gamma = \frac{1}{12} - \xi, \ P = \frac{1}{\epsilon^2} \frac{d}{dt}.$$

From the form of the characteristic equation $p^2 + 2\eta p + \eta^2 + \gamma^2 = 0$, it is easy to conclude that the system is asymptotically stable when the oscillator's damping is positive, that is, $\eta > 0$.

The susceptibility averaged over the oscillation period of pumping is derived from (3) as the ratio y_0/x_0 and can be expressed by the operator

$$\alpha(p) = 1 - \frac{1}{4} \frac{\gamma}{p^2 + 2hp + \eta^2 + \gamma^2}. \tag{4}$$

The static susceptibility becomes negative when $\gamma/(h^2+\gamma^2) > 4$ and reaches the maximum of its absolute value $\alpha_{(-)} = \frac{1}{8}\eta - 1$ at the detuning $\xi = \frac{1}{12} - \eta \ (\eta < \frac{1}{8})$. The operator (4) is associated in this case with the averaged response to the Heaviside function, that is, with the transition process

$$A(\tau) = 1 - \frac{1}{8\eta}[1 - e^{-\eta\tau}(\cos \eta\tau + \sin \eta\tau)], \tag{5}$$

where $\tau = \epsilon^2 t$ is the 'slow' time. The duration of this process increases with the increase of the desired negative susceptibility and the decrease of the square of the pumping amplitude ϵ^2.

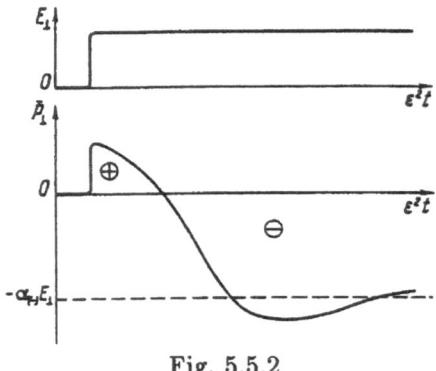

Fig. 5.5.2

When the transverse field is introduced adiabatically (Figure 5.5.2), the following process occurs. At first, in view of (5), the transverse dipole polarization (averaged over a period) will be oriented in the direction of the field, but then the inversely-directed component of polarization will increase with the time constant $\theta = 1/\epsilon^2 \eta$. At the end of the transition process, the resulting averaged polarization becomes negative if $\eta < 1/8$. Obviously, the energy of transverse polarization is drawn from the pumping source, that is, a preliminarily excited non-linear oscillator that has the second harmonic in its oscillation spectrum. Energy exchange between the two oscillators is carried out via the non-linear connection between them and is generally bilateral; it is reversible at $\eta = 0$. In this case we have, instead of (1), the weakly non-linear system of equations

$$\ddot{y} + (1 + 2\epsilon^2\xi)y + \epsilon yz = x_0, \quad \ddot{z} + (1 + 2\epsilon^2\xi)z + \epsilon\beta z^2 = -\epsilon y^2/2.$$

When the absolute value of y^2 is small enough, the second equation can be solved independently, the spectrum of $z(t)$ possesses the first and the second harmonics, and we arrive at the result above with an almost independent parametric pumping.

When non-linearly connected stationary or travelling waves are employed as oscillations, the negative susceptibility of the system will reveal itself with respect to the stationary or travelling waves of the external field, and hence distributed control can be realized (Samoilenko, 1974).

As is clear from the examples above, the state of negative susceptibility occurs under certain conditions in dynamic systems of various mathematical descriptions and physical natures. However, the following general features of this phenomenon can be formulated:

1. It appears when the system has an energy accumulator possessing in its non-perturbed state the maximum potential energy in an even number of coordinates.

2. The stability of the system in a state of negative susceptibility requires the employment of explicit or implicit (as in the last example) generalized gyroscopic forces.

3. Asymptotic stability in the absence of inertial forces can be achieved using the generalized forces of the non-conservative position-sensitive type.

4. Under a sufficiently slow external influence, the energy is transferred from the system with negative susceptibility to the external perturbation source if the perturbation increases and in the reverse direction when the perturbation diminishes. It is worthwhile to use this phenomenon for the stabilization of distributed and fast processes when the impedance method is applied.

5.6. Systems with Stored Energy

It has been noted above that in a number of cases, the construction of control systems requires rapid distributed negative feedback with amplification (Butkovskiy, 1975, 1977). The possibility of producing such a feedback is considered in Samoilenko (1968, 1978b) on the basis of the phenomenon of negative generalized susceptibility in dynamic systems with either stored internal energy or external parametric excitation.

The properties of systems with stored energy will be discussed using mechanical models as illustrations. Obviously, the part of stored energy which does not depend on the external (environmental) conditions is inessential for the characteristics of both resistivity and susceptibility. Therefore 'stored energy' in this context refers only to the component of total energy that does not depend on external coordinates or forces and is at its maximum in the non-perturbed state of the system.

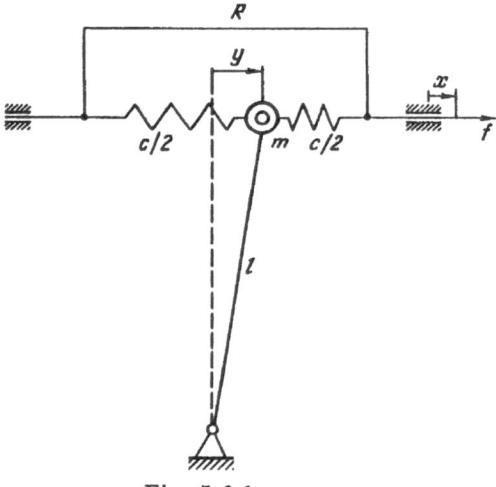

Fig. 5.6.1

The astatic pendulum shown in Figure 5.6.1 is one of the simplest devices with stored potential energy (Nikolai, 1952). At small displacements from the equilibrium position, its motion can be described by the equations

$$m\left(p^2 - \frac{g}{l}\right)y = c(x - y) = f,$$

where m is the mass of the pendulum, l is its length, g is the acceleration due to gravity, c is the resistivity (rigidity) of the springs, x and y are the displacements of the pendulum and the frame R (we neglect the mass of the frame), f is an external force, and $p = d/dt$ (where t is time). Here x is the external coordinate and y is the internal one. Eliminating y and expressing f in terms of x, we obtain

$$f = c(p)x, \quad c(p) = -c_{(-)}\frac{1 - \tau^2 p^2}{1 + T^2 p^2},$$

$$c_{(-)} = c\left(\frac{cl}{mg} - 1\right)^{-1}, \quad \tau^2 = \frac{l}{g}, \quad T^2 = \frac{\tau^2 c_{(-)}}{c}.$$

Here $c(p)$ is the resistivity operator, $c_{(-)}$ is the absolute value of negative static resistivity, and τ and T are time constants.

It is easy to verify that the operator $c(p)$ meets the conditions of physical realizability formulated in §5.1. In particular, the ratio of the coefficients at greater powers of p in the numerator and the denominator equals $+1$. The condition for neutral stability of internal motion is satisfied at a fixed external coordinate x if $c > mg/l$, that is, if the resistivity of the springs is large enough. Conversely, this device is unstable at a fixed external force and a negative static resistivity.

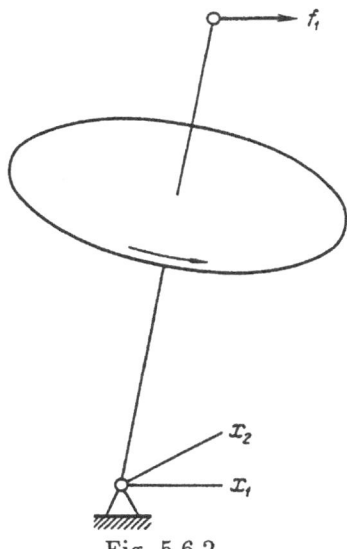

Fig. 5.6.2

An example of a device possessing a negative static susceptibility and stable at a fixed external force is a heavy symmetrical top with a fixed support point (Figure 5.6.2). Linearized equations of the motion of axis of the top induced by a transverse force f can be represented (Samoilenko, 1978a) in the form

$$(mp^2 - c_{(-)})x_1 - \gamma p x_2 = f_1,$$
$$\gamma p x_1 + (mp^2 - c_{(-)})x_2 = 0,$$
$$(1)$$

where m is the equivalent mass, $c_{(-)}$ is the absolute value of negative resistivity, γ is the gyroscopic parameter, and x_1 and x_2 are the transverse coordinates of the point where the force f is applied in the direction of the axis x_1. Eliminating the internal coordinate x_2, we obtain the dependence of $x_1 = \alpha(p)f_1$ on the susceptibility operator

$$\alpha(p) = -\alpha_{(-)}\frac{1 - \theta^2 p^2}{1 + 2\eta^2\theta^2 p^2 + \theta^4 p^4},$$
$$(2)$$

where

$$\alpha_{(-)} = c_{(-)}^{-1}, \quad \eta^2 = \gamma^2/2mc_{(-)} - 1, \quad \theta = \sqrt{m/c_{(-)}}.$$

Let us verify that the conditions of physical realizability are observed.

a) the degree of the numerator of the operator (2) is less than that of the denominator by two;

b) the roots of the denominator $\theta^{\pm}p_{\pm} = \pm i\sqrt{\eta^2 \pm \sqrt{\eta^4 - 1}}$ do not equal the roots of the numerator $\theta p_{\pm} = \pm 1$, if $\gamma \neq 0$;

c) the ratio of the coefficients of the terms of greatest degree in the numerator and the denominator equals $m^{-1}c_{(-)}$ and is always positive.

The condition of neutral stability under a fixed external force is satisfied if $\gamma > 2(mc_{(-)})^{1/2}$, that is, if the gyroscopic forces are great enough. By contrast, if the coordinate x_1 is fixed by external restraints, then the system becomes unstable. This follows formally from the fact that the equation $\alpha(p)f_1 = x_1 \equiv 0$ possesses an unstable solution with the increment $p = (c_{(-)}/m)^{1/2}$. The static susceptibility can be deduced from (2) for $p = 0$ and it has a negative value of $-c_{(-)}^{-1}$, which is the reciprocal of the static resistivity. As $c_{(-)}$ diminishes, the absolute value of the negative susceptibility can become arbitrarily large.

In the general case, (2) can be reduced to the form

$$\alpha(p) = -\frac{1}{1 - \epsilon}\frac{\alpha_{(-)}}{\theta_0^2 p^2 + 1} + \frac{\epsilon}{1 - \epsilon}\frac{\alpha_{(-)}}{\epsilon^2\theta_0^2 p^2 + 1},$$
$$(3)$$

where $\epsilon = (\eta^2 + \sqrt{\eta^4 - 1})^{-1}$, $\theta_0^2 = \epsilon^{-1}\theta^2$. If $m \to 0$ for $\tau = \gamma/c_{(-)} =$ const, then $\epsilon \to 0$ and the operator (3) can be replaced in the limit by the simpler expression

$$\alpha(p) = -[c_{(-)}(\tau^2 p^2 + 1)]^{-1}.$$
$$(4)$$

However, it should be borne in mind that the idealized operator (4) does not meet the condition c) of physical realizability.

The operator (3) meets every condition of physical realizability and the condition of stability. For instance, the condition c) holds because of the second term in (3). This is why there is a positive peak in the initial part of the transition process in response to the Heaviside function at the input. The transition process as a whole can be characterized as the occurrence of non-damped oscillations about the averaged negative value of the x_1 coordinate.

Damping can be introduced into (1) by using negative friction forces, the implementation of which is not considered here. There are also other possibilities to synthesize stable systems with negative susceptibility and damped free oscillations. Some of the general features of such systems deduced from an analysis of the structure of forces within the framework of a linear stationary phenomenological model are formulated in Samoilenko (1978a).

As to the role of stored energy, the following can be said. In considering the motion of an axisymmetric top in Eulerian coordinates, the angle of pure rotation is a cyclic variable. In both this case and the more general one, it is convenient to eliminate the cyclic coordinates (Merkin, 1974) by using the Routhian function. The system resulting from the elimination of cyclic coordinates is said to be *reduced*. At small deviations of the position coordinates from the equilibrium point ($q_k = 0$, $\dot{q}_k = 0$), the system's linearization yields the equations of perturbed motion with constant coefficients

$$\sum_{j=1}^{S}(a_{kj}\ddot{x}_j + g_{kj}\dot{x}_j + c_{kj}x_j) = X_k, \quad k = 1,\ldots,S,$$

where x_j and X_k are the perturbations of the generalized coordinates and forces, and a_{kj}, g_{kj}, c_{kj} are the matrices of the inertial, gyroscopic, and potential forces respectively.

Since $g_{jk} = -g_{kj}$, the total power of the gyroscopic forces equals zero. For this reason, under small perturbations the energy can only be stored in the form of the potential and kinetic energies of the position coordinates. However, for static susceptibility, only the potential forces play an essential role, because the inertial forces in the mode of steady oscillations vanish on the average. It follows that in the case in question the kinetic energy stored in cyclic motions is only utilized to produce static susceptibility because the kinetic energy changes the form of the potential function of the reduced system owing to the generalized centrifugal forces. Note that this kind of force is absent in the axially symmetric top. Another function of the cyclic motion is to produce gyroscopic forces stabilizing the equilibrium when the number of negative eigenvalues is even in the matrix of potential forces (Merkin, 1974). This was used in the model of a top with negative susceptibility.

Thus when using systems whose perturbed motion is described by stationary linear equations, one should utilize stored energy and stabilizing gyroscopic forces in order to produce negative susceptibility. In the case of non-stationary equations of perturbed motion, the analysis of conditions for negative susceptibility calls for other approaches, which will be considered below.

5.7. Static Susceptibility of Adiabatically Invariant Control Systems

Automatic control systems realized on the basis of the negative susceptibility phenomenon can possess, in a number of practical cases, *adiabatic invariance*, that is, the variables that do not depend on time but depend on slowly varying parameters and stored energy. For instance, Hamiltonian systems with separable variables can have adiabatic invariants such as the action integrals

$$I_i = (2\pi)^{-1} \oint p_i dq_i,$$

where p_i, q_i are the momenta and coordinates of the subsystems into which the system falls after the separation of variables, the integrals being taken over particular cycles of conventionally periodic motion (Landau and Lifshiz, 1965). In thermally isolated statistical ensembles, the adiabatic invariant is the entropy S. If the adiabatically invariant systems possess negative susceptibility in a state with stored energy, they can be used effectively to produce a negative feedback in control systems for rapid processes with both distributed and lumped parameters.

Statistical physics proves (Landau and Lifshits, 1976) that in adiabatically invariant classical or quantum statistical ensembles the following formula holds:

$$\overline{\frac{\partial \hat{H}}{\partial \lambda}} = \left(\frac{\partial E}{\partial \lambda}\right)_S, \tag{1}$$

where $\widehat{H}$ is the Hamiltonian, E is the average energy, λ is a slowly varying parameter, and the bar means total averaging, both quantum-mechanical and over the ensemble. If the time scales of 'rapid' and 'slow' motions are essentially different and the system is non-degenerate, the averaging over the ensemble can be substituted by time-averaging.

Let a system have an adiabatic invariant I. In the presence of external forces f_k, $k = 1, \ldots, m$, the average energy of the system is less than its initial value E_0 by the value of the work performed to overcome these forces:

$$E(f_1, \ldots, f_m, I) = E_0 - \sum_{k=1}^{m} \bar{q}_k f_k. \tag{2}$$

By differentiating keeping I constant, we can find the average values of the external coordinates $\bar{q}_k = -(\partial E / \partial f_k)_I$ and the coefficients of the averaged linear susceptibility

$$\alpha_{ik} = \left(\frac{\partial \bar{q}_k}{\partial f_j} \right)_I = -\left(\frac{\partial^2 E}{\partial f_j \partial f_k} \right)_I. \tag{3}$$

Hence the energy of the adiabatically invariant system with negative susceptibility, as a function of external forces, reaches a local minimum when (3) becomes zero.

The adiabatic invariant can be regarded as a function of the energy and the external forces:

$$I = I(E, f_1, \ldots, f_m) = \text{const.}$$

Differentiating this expression with respect to the components of the external force, we obtain

$$q_k = \frac{\partial I / \partial f_k}{\partial I / \partial E},$$

$$\alpha_{jk} = \frac{(\partial^2 I / \partial f_j \partial f_k)(\partial I / \partial E) - (\partial^2 I / \partial f_j \partial E)(\partial I / \partial f_k)}{(\partial I / \partial E)^2}. \tag{4}$$

Here the derivatives are to be calculated at the point $f_k = 0$, $k = 1, \ldots, m$.

Using this approach, let us show that a flat rotator (Figure 5.7.1) possesses a negative susceptibility with respect to a slowly varying gravity field. The energy integral for such a system can be expressed in Cartesian coordinates as

$$\frac{p_\tau^2}{2m} - fq = E,$$

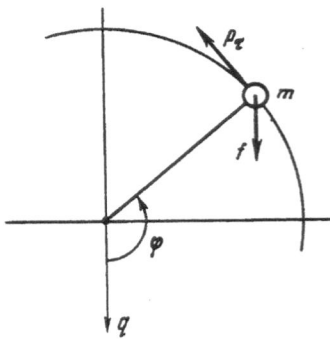

Fig. 5.7.1

where m is the mass, p_τ is the tangential momentum, q is the vertical coordinate, and $f = mg$ is the force of gravity, which in this example is a slow
function of time.

It is more convenient for the analysis to express the energy in terms of the
polar angle ϕ measured from the direction of the q-axis:

$$\frac{p_\phi^2}{2J} - fr\cos\phi = E. \tag{5}$$

Here $p_\phi = p_\tau r$ is the angular momentum and $J = mr^2$ is the moment of inertia.
The adiabatic invariant I in this case can be expressed in terms of E and f as
follows:

$$I = (2\pi)^{-1} \oint p_\phi d\phi = \pi^{-1} \int_0^\pi \sqrt{2J(E + fr\cos\phi)}\,d\phi.$$

In order to determine the linear susceptibility, it is sufficient to assume
that $fr \ll E$ and represent I to within f^2 as follows:

$$I \simeq \sqrt{2JE}\left(1 - \frac{f^2 r^2}{16E^2}\right).$$

Using this expression, we find according to (4) that $\bar{q} = -r^2 f/4E$, $\alpha = -r^2/4E$.

Since $E > 0$, the static susceptibility α proves to be negative. This is accounted for by the fact that while the rotator mass moves upwards, its rotation
decelerates, and while it moves downwards the rotation accelerates. Therefore,
the rotator mass is in the upper half of the trajectory longer than in the lower
half. In view of the uneven rotation the position of the rotator's mass centre

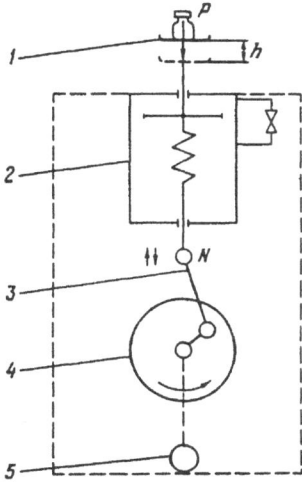

Fig. 5.7.2

averaged over the trajectory is displaced upwards with respect to the centre of trajectory, that is, in the opposite direction to the force applied.

The mechanism represented in Figure 7.5.2 can demonstrate a rather unusual feature based on the negative susceptibility phenomenon in the rotator. When the weight P is put into the cup 1, the weight does not go down as it might be expected but rises to a height h. This occurs because the weight is applied via the oscillation damper 2 and the crank mechanism 3 to the flywheel 4, resulting in a *non-uniform* rotation of the flywheel. The point N at which the crank mechanism is connected with the damper rod performs vertical oscillations and is displaced upwards on the average, and therefore the weight P goes up. Rapid oscillations are eliminated by the damper and may be unnoticed by the observer. The energy stored by the rotating flywheel is mainly used to demonstrate the effect. A small loss of energy is compensated for by a low-power motor 5.

We can connect kinematically this mechanism with an unstable system, for example, the astatic pendulum shown in Figure 5.6.1, and then the mechanism can be used as the automatic controller of the equilibrium. Phenomenologically its role is to render the total resistivity positive. When mechanical elements are connected consecutively, their displacements and therefore their susceptibilities are summed. The total resistivity of the springs possessing resistivity c and the controller possessing negative susceptibility $-\alpha_{(-)}$ equals

$$c_\Sigma = (c^{-1} - \alpha_{(-)})^{-1}.$$

As $\alpha_{(-)} \rightarrow c^{-1} - 0$, the total resistivity may become arbitrarily large and sufficient to stabilize the astatic pendulum.

Naturally, there can be many versions of the practical realization of the principle outlined above. For instance, kinetic energy can be stored in translational rather than rotational motion of bodies in a field of forces with a periodic potential. The form of the energy equation (5) will not change, but the variable ϕ is to be interpreted as the longitudinal coordinate of the translational motion. If there is a periodic field and there are many bodies moving in it and weakly interacting both with each other and with the walls of the channel (for example, a thermally isolated compressible fluid or an electron gas), it may be advisable to consider a statistical ensemble, whose adiabatic invariant is the entropy.

If we assume in (5) that $fr = HP_m$, where H is the strength of the magnetic field and P_m is the magnetic moment of the rotator, it follows from (4) and (3) that the static susceptibility of the magnetic rotator is negative. This is the physical explanation of the stability of a magnet levitating in the field of a rotating magnet studied in Van der Heide (1976), where stability diagrams for the solutions of the Mathieu differential equation are used.

5.8. Conditions for Negative Static Susceptibility in Quantum Systems

We have already remarked that the transition to a state with negative susceptibility can be regarded as an objective of quantum control. The static susceptibility of quantum systems can be considered as the limit value of susceptibility with respect to a periodic field whose frequency vanishes. One can use here the well-developed apparatus of the quantum theory of linear susceptibility (Blokhintsev 1981a, 1981b). For the sake of definiteness, we shall take the generalized force to be the amplitude $E(\omega)$ of an electric field varying with time according to a harmonic law with the frequency ω and the external generalized coordinate to be the amplitude $P(\omega)$ of an electric dipole moment.

The electric susceptibility $\alpha_n(\omega)$ in the linear relation $P(\omega) = \alpha_n(\omega)E(\omega)$ for the state with a definite value of the energy E_n is obtained from the formula

$$\alpha_n(\omega) = \frac{e^2}{m_e} \sum_k \frac{f_{kn}}{\omega_{kn}^2 - \omega^2}. \tag{1}$$

Here e and m_e are the charge and mass of the electron, $\omega_{kn} = (E_k - E_n)/\hbar$ is the transition frequency, and E_k are stationary energy levels. The quantity

$$f_{kn} = \frac{2m_e}{\hbar} \omega_{kn} |x_{kn}|^2,$$

where $\hbar$ is Planck's constant, is called the *force* of the oscillator with frequency ω_{kn}. The matrix elements of the operator $\hat{x}$ of the coordinate x in the direction of the external field are defined as

$$x_{kn} = \langle k|\hat{x}|n\rangle. \tag{2}$$

The left $\langle k|$ and the right $|n\rangle$ orthonormal eigenfunctions (using the Dirac brackets) of the non-perturbed Hamiltonian $\hat{H}_0$ can be found from the equations

$$\hat{H}_0|n\rangle = E_n|n\rangle, \quad \langle k|\hat{H}_0 = \langle k|E_k. \tag{3}$$

The static susceptibility $\alpha_n = \alpha_n(0)$ can be obtained from (1) at $\omega = 0$:

$$\alpha_n = 2e^2 \sum_{k \neq n} \frac{|x_{kn}|^2}{E_k - E_n}. \tag{4}$$

This allows us to formulate the general condition for the occurrence of negative static susceptibility of a quantum system in a state with energy E_n:

$$\sum_{k<n} \frac{|x_{kn}|^2}{E_k - E_n} > \sum_{k>n} \frac{|x_{kn}|^2}{E_k - E_n}. \tag{5}$$

As usual, the states are numbered here in the ascending order of energy levels. Obviously, if the system is in its lowest energy state, the condition (5) cannot hold. In other words, negative susceptibility is only possible when the system is in an excited state, that is, when it possesses some stored energy.

Let us show that the electric susceptibility of a flat rotator, namely an electron on a circular orbit of constant radius r, can be negative. The non-perturbed Hamiltonian expressed in terms of the polar coordinate ϕ measured from the axis has the following form:

$$\hat{H}_0 = -\frac{\hbar^2}{2J} \frac{\partial^2}{\partial\phi^2},$$

where $J = mr^2$ is the electron's moment of inertia; equations (3) yield

$$|n\rangle = \frac{1}{\sqrt{2\pi}} e^{in\phi}, \quad \langle k| = \frac{1}{\sqrt{2\pi}} e^{-ik\phi}, \quad E_n = \frac{\hbar n^2}{2J}. \quad (6)$$

Since $x = r\cos\phi$, the operator $\hat{x}$ in this representation is the operator of multiplication by the variable x which depends on ϕ. Taking this into account we can define in view of (2):

$$x_{kn} = \frac{r}{2\pi} \int_0^{2\pi} e^{i(n-k)\phi} \cos\phi \, d\phi = \frac{r}{2}(\delta_{k,n+1} + \delta_{k,n-1}).$$

Substituting this into (4), we find the static susceptibility of the rotator in the state n:

$$\alpha_n = \frac{e^2 r^4 m_e}{\hbar^2} \frac{1}{1 - 4n^2} = \frac{p^2 J}{\hbar^2} \frac{1}{1 - 4n^2}, \quad (7)$$

where P is the dipole moment of the rotator and J is its moment of inertia. This suggests that if the rotator is not in its ground (lowest) energy state, its susceptibility is negative, the greatest value of the susceptibility corresponding to $n = 1$. Similar calculations for a quantum oscillator with Hamiltonian

$$\widehat{H}_0 = \frac{\hat{p}^2}{2m_e} + \frac{m_e \omega_0 \hat{x}^2}{2},$$

where $\hat{p}, \hat{x}$ are the operators of the coordinate and the momentum, m_e is the mass and ω_0 is the oscillator's frequency, yield the following results:

$$|x_{kn}|^2 = \frac{(k\delta_{k-1,n} + n\delta_{n-1,k})\hbar}{2m_e \omega_0},$$

$$E_n = \left(n + \tfrac{1}{2}\right)\hbar\omega_0,$$

$$\alpha_n = 2e^2 \sum \frac{|x_{kn}|^2}{E_k - E_n} = \frac{e^2}{m_e \omega_0^2} > 0.$$

The last expression eliminates the possibility of negative susceptibility of this oscillator either in the ground state or in an excited state. It can therefore be concluded that a state of negative susceptibility is not possible for all physically real systems.

An example of a flat rotator interacting with the external field is a molecule possessing a dipole moment. If the momentum of the field of circular polarization starts rotating the molecule around an axis perpendicular to the direction of the dipole moment, the molecule will possess negative susceptibilities in the directions that are transverse to the rotation axis.

If at least one of the coefficients f_{kn} is negative, this leads to an anomalous frequency dependence of the refractive index, the phenomenon being called *negative dispersion* (Blokhintsev, 1981a). As to negative static susceptibility, it is generally insufficient to produce negative dispersion in the vicinity of a local resonance: inequality (5) must hold for all resonance transitions. In particular, this is realizable in systems with negative temperature, such as, for instance, the adiabatically demagnetized spin gas (Landau and Lifshits, 1976) or in excited two-level systems (one of them will be discussed below).

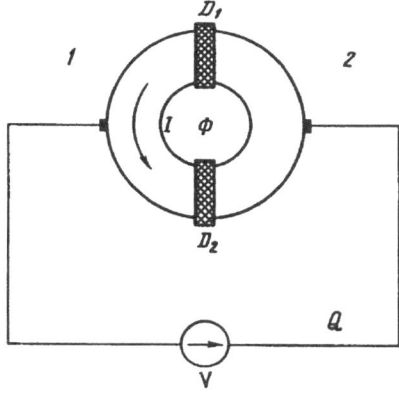

Fig. 5.8.1

Suppose that there is a superconducting ring (Figure 5.8.1) which has two diametrically opposite narrow dielectric slits D_1 and D_2 known as *Josephson junctions*. Prior to cooling the metal to achieve the superconducting state, a magnetic flux $\Phi_i = n\pi\hbar/e$ is introduced into the ring. This flux is maintained after the transition to superconductivity. Then the direction of the external field that created the initial flux Φ_i is reversed, the result being the appearance of an electric current and stored magnetic energy $LI^2/2 = \Phi_e^2/2L$, where Φ_e is the flux of the inverse field and L is the self-induction of the ring. Then if the half-rings separated by the Josephson junctions are connected adiabatically to an external source of voltage V, an electric charge Q will be passing in the direction opposite to the voltage applied, and this will be evidence of negative electric capacity.

Using the elementary theory of Josephson junctions (Feynman et al., 1963), we perform some calculations. We denote by ψ_1 and ψ_2 the probability amplitudes that a Cooper pair of electrons is in the half-rings 1 and 2 respectively. Assume the energies of the Josephson junctions to be identical and denote them by K. Suppose a Cooper pair with the charge $q = -2e$ occurs in the half-ring 1 at time t. In view of the influence of the magnetic field with vector potential A on the transition phases (Feynman et al., 1963), we can obtain the amplitude of the probability that a Cooper pair occurs in the semi-ring 2 at time $t + dt$:

$$d\psi_2 = -\left(\frac{i}{\hbar}\right)\left\{K\exp\left[\left(-\frac{iq}{\hbar}\right)\int_1^2 A\,ds\right] + K\exp\left[\left(-\frac{iq}{\hbar}\right)\int_1^2 (-A)ds\right]\right\}dt.$$

The integrals here are to be taken over the two paths of transition between the semi-rings. Because of the symmetry of transitions,

$$\int_1^2 A\,ds = \frac{1}{2}\oint A\,ds = \frac{\Phi_i}{2}.$$

Therefore $d\psi_2 = -(i/\hbar)2k\cos\phi\,dt$, $\phi = e\Phi_i/\hbar$. This result allows us to describe the dynamics of the transitions by the following system of equations:

$$\begin{aligned} i\hbar\dot{\psi}_1 &= \left(-\frac{qV}{2}\right)\psi_1 + 2k\cos\phi\psi_2, \\ i\hbar\dot{\psi}_2 &= 2k\cos\phi\psi_1 + \left(-\frac{qV}{2}\right)\psi_2. \end{aligned} \tag{8}$$

The stationary solutions of these equations contain the phase factor $\exp(-iE^{(\pm)}t/\hbar)$, where the possible values of the energy $E^{(\pm)}$ are found from the characteristic equation of the system (8) and represented for small V/k in the form

$$E^{(\pm)} \simeq \pm(2k\cos\phi + q^2 V^2/16k\cos\phi).$$

The energy $E^{(\pm)}$ is plotted against V in Figure 5.8.2. Typically the upper curve has its minimum at $V = 0$. Here the susceptibility defined for a Cooper pair in compliance with (7.3) as

$$\alpha = -\frac{\partial^2 E}{\partial V^2} = -\frac{q^2}{8k\cos\phi},$$

becomes negative for $N/2$ pairs: $\alpha = -N^2 q^2/32k\cos\phi$. An electric charge

$$-eN = -Q = -\alpha N = N^2 q^2 V/32k\cos\phi$$

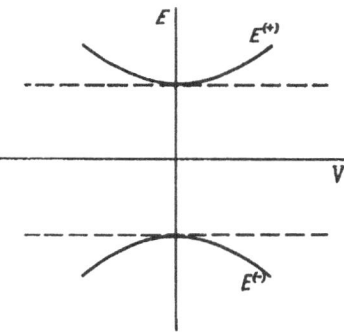

Fig. 5.8.2

will pass in the opposite direction to the applied voltage V. In order to achieve a state with the quantity of stored energy $E^{(+)} = \Phi_e^2/2L$, it is necessary to have the appropriate value of the reversely directed external flux Φ_e.

Equations (8) are typical for any two-level systems with transition energy $W = 2k\cos\phi$. Therefore properties similar to the ones indicated above will also be possessed by any other system with a perturbing potential of the form $-qV$ while the system is at its upper energy level.

The use of the negative susceptibility phenomenon opens up new fundamental vistas of realizing rapid negative feedback on structurally uniform elements. It can find application in distributed control systems for fast processes. However, it is worth bearing in mind that the synthesis of distributed structures and quasicontinuous control media (Samoilenko, 1968) using lumped parameter elements with negative susceptibility brings about specific problems of the stability of the system as a whole. The interaction between these elements may disturb the stability and lead to the formation of the 'domain structure' (Landau and Lifshits, 1959).

The stability of a linear stationary control medium, in compliance with the general theory of linear systems, can be investigated examining the distribution of the poles of the resulting transition function — the linear susceptibility operator in this case. Given the simplest cubic lattice of the control medium, we can find its susceptibility from the susceptibility of the 'atom' using the concept of the 'internal effective field' introduced in the electrodynamics of continuous media (Landau and Lifshits, 1959) as the field producing a positive feedback in the system. As to the analysis of other more complicated structures, it is advantageous to apply the well-developed methods of the structural theory of distributed systems (Butkovskiy, 1977).

There are many cases in which the above-mentioned approach can be used for a direct analysis of the susceptibility of distributed systems without considering the susceptibilities of individual structural elements. For this the amplitudes of eigenvibrations can be regarded as the generalized coordinates, and the methods of secondary quantization can be employed in the quantum case (Blokhintsev, 1981a). The behaviour of negative susceptibility elements within complicated systems calls for further study.

CHAPTER 6

Negative Susceptibility in Parametrically Induced Magnetics

6.1. Induced Superdiamagnetism and its Application to Distributed Control

This chapter will show that pulse remagnetization and synchronous longitudinal pumping can, under certain conditions, transfer gyromagnetic media into a superdiamagnetic state that typically possesses negative susceptibility with respect to quasistatic perturbations. We shall also discuss the trends in the employment of these phenomena for distributed electromagnetic control.

It has already been noted that by raising the requirements to include fast performance and space resolution of distributed systems, one introduces the problem of searching for new methods and technical means of distributed automatic control for rapid processes in continuous media. In a number of cases, the problems of automatic stabilization of plasma and particularly of relativistic electron or electron-ion rings cannot be solved by using amplifiers and other devices with lumped parameters. This situation arises because the increments of unstable perturbations in such processes are very large (10^6–$10^9\,S^{-1}$), and even a very small delay can eliminate the chance of achieving stability by means of automatic control of electromagnetic fields.

By analogy with the methods employed to amplify or transform wide-band signals in radioelectronics (Alekseev, 1968; Pantell and Puthoff, 1969; Tarasov, 1976), the techniques of automatic control of the above processes should first focus on the following problems:

a) to sustain distributed interaction between the process and the control device, both possessing distributed parameters;

b) to employ parametric field amplification and transformation.

An analysis of the general ideas of distributed electromagnetic control (Samoilenko, 1968) shows that an 'ideal' distributed controller of a magnetic field can be a control medium with negative magnetic permeability, that is, a magnetic possessing a negative susceptibility greater than unity in absolute value. As to stability, this medium should also possess an appropriate space dispersion (Samoilenko, 1968).

Media with negative magnetic susceptibility are commonly called diamagnetics. The free energy of any magnetic reaches its minimum at the state of thermodynamic equilibrium; in the vicinity of the equilibrium it is a bilinear form of the induction B and the magnetic field H. This is why the magnetic permeability μ of the media in thermodynamical equilibrium, defined as

$$\left(\frac{\partial B}{\partial H}\right)_{\xi,T} = \mu,$$

where ξ is the chemical potential and T is temperature, is an essentially positive quantity. Diamagnetics are an exception. Therefore a state with a negative μ is either a metastable state of a continuous medium or a state that is stable but parametrically excitable. We call a state *superdiamagnetic* if μ is less than zero and the magnetic susceptibility is not only less than zero (as in diamagnetics) but also greater than unity in absolute value. Usually, the term 'superdiamagnetism' is used (Vonsovskii, 1971) as a synonym of 'marginal' diamagnetism at $\mu = 0$ in systems in thermodynamical equilibrium. It is advantageous to extend this term to the region where $\mu < 0$ in order to cater for the non-equilibrium systems discussed below.

Note that in gyromagnetic media, owing to magnetic resonance, the real part of the transverse magnetic susceptibility

$$\mu_1' = 1 + \chi_0 \frac{1 - \nu^2}{(1 - \nu^2)^2 + 4\alpha^2\nu^2},$$

where

$$\chi_0 = \frac{M_0}{H_0}, \quad \nu = \frac{\omega}{\gamma H_0},$$

γ is the magnetomechanical ratio, M_0 is the initial (spontaneous) magnetization and H_0 is the magnetizing field, which becomes negative (at low enough damping parameter α) in the range of frequencies ω of the external field above the resonance frequency $\omega = \gamma H_0$, that is, when $\gamma > 1$. This negative permeability, which is useful for distributed control in periodic fields, can be regarded as a trivial case of superdiamagnetism of resonance type. Below we shall discuss the possibilities of obtaining a wide-band superdiamagnetism that reveals itself in static or slow enough quasistatic fields whose spectrum abuts the zero frequency.

In assessing the realizability of a superdiamagnetic state, we have to take into account, besides the requirements of thermodynamic stability, the demands of macroscopic and domain stability. Here an essential role is played by the space-time dispersion properties of the medium and by the boundary conditions describing its interaction with the environment. An investigation of these factors can be regarded as an independent, extraneous problem of distributed control.

Below we are going to discuss two possible methods of obtaining a parametrically induced superdiamagnetism. One of them is based on longitudinal pulse magnetization and the other on the synchronous periodic modulation of the field magnetizing a gyromagnetic medium. The stability of the state is analyzed under the assumption of a weak connection with the environment, when the external field can be regarded as being independent of the magnetization. The boundary conditions for the gyromagnetic medium within the surface of an ellipsoid are accounted for by introducing demagnetizing factors characterizing the shape anisotropy of the specimen.

6.2. Superdiamagnetic States in Inversely Magnetized Ferromagnetic Media

The evolution of the magnetization vector M in the process of remagnetization of a ferromagnetic medium by the effective magnetic field H_Σ is described by the Landau-Lifshits equation (Landau and Lifshits, 1935)

$$\frac{dM}{dt} = -\gamma M \times H_\Sigma + R, \tag{1}$$

where R is a dissipation term whose concrete form is defined by the mechanism of dissipation depending on the law of magnetization and the microstructure of the medium (Osborne, 1945; Mikhaelyan, 1963; Landau and Lifshits, 1935).

The magnetomechanical ratio is expressed in SI units as

$$\gamma = \frac{g}{2}\frac{|e|}{m}\mu_0,$$

where g is the spectroscopic splitting factor (usually $g \approx 2$), e and m are the charge and mass of the electron and μ_0 is the magnetic constant.

Let us consider an ellipsoidal specimen made of a monocrystal ferrite with a cubic lattice whose principal directions $\begin{bmatrix} 1\,0\,0 \end{bmatrix}$, $\begin{bmatrix} 0\,1\,0 \end{bmatrix}$, $\begin{bmatrix} 0\,0\,1 \end{bmatrix}$ and the anisotropy axes of the specimen's shape are oriented along the unit vectors e_i, $i = 1, 2, 3$, in the Cartesian system of coordinates. The effective field

$$H_\Sigma = H - H_f - H_a \tag{2}$$

consists of the external magnetizing field

$$H = H_1 e_1 + H_2 e_2 + H_3 e_3, \tag{3}$$

a demagnetizing field of anisotropy

$$H_f = N_1 M_1 e_1 + N_2 M_2 e_2 + N_3 M_3 e_3, \tag{4}$$

where the N_i are the demagnetizing factors, and the field of crystallographic anisotropy (Kukharkin, 1969)

$$H_a \simeq -\frac{2K_1}{\mu_0 M^4}(M_1^3 e_1 + M_2^3 e_2 + M_3^3 e_3) \quad (M = |M|), \tag{5}$$

where K_1 is the first constant of crystallographic anisotropy (usually $K_1 < 0$).

Assume that at the beginning the magnetic field H is great enough to produce a homogeneous magnetization M of the whole specimen without the formation of domains and that H is oriented along the unit vector e_3. The dissipation term R vanishes at the termination of the relaxation processes. This state becomes equilibrium because the vector product in (1) equals zero as a result of the parallel orientation of M and H when $M \times H_\Sigma = 0$, $M \cdot H_\Sigma > 0$.

Now if we perform a rapid longitudinal (that is, along the initial direction of H) remagnetization of the specimen for a small enough interval $T_0 \ll T_\parallel$, where $T_\parallel$ is the relaxation time of the longitudinal component of magnetization, then during a period of order $T_\parallel$ the ferrite will be in the state of inverse magnetization, in which M and H are antiparallel. At both the initial state and this one, $M \times H_\Sigma = 0$, but $R \neq 0$, and therefore there is no full equilibrium, and the state can be said to be metastable. The specimen relaxes to the initial state during a period of order $T_\parallel$.

Let us describe the properties of an inversely magnetized ferrite occurring after a rapid (pulse) remagnetization for an interval that is small compared with $T_\parallel$. Owing to remagnetization, the ferrite stores an energy whose density is $(\mu_0/2)M \cdot H$ and which can be exchanged with the external perturbing field. Depending on the sign of the magnetic permeability μ, as the energy density $(\mu_0/2)H_2$ of the external field increases, the energy density $(\mu_0\mu/2)H_2$ within the magnetic can either increase ($\mu > 0$) or decrease ($\mu < 0$). In the latter case, the superdiamagnetic medium releases energy and builds up the energy density in the environment, therefore exhibiting the functions of a controller of the magnetic pressure.

Since the state of the ferrite relaxes to the equilibrium after the pulse remagnetization, the magnetic permeability tensor varies with time during a characteristic interval of order $T_\parallel$, which we can assume to be considerably greater than the period of free precession of the magnetic moment. We can introduce the notion of the 'instantaneous tensor of magnetic permeability' here. In order to define it, we have to give a concrete form to the dissipation term in equation (1). It is usually given for the state where $M \cdot H > 0$. Therefore, as is appropriate for the state of inverse magnetization, we shall use the well-known definition of relaxation substantiated in Skrotsky and Kurbatov (1961). According to this definition, the components of magnetization relax to the equilibrium orientation determined by the total magnetic field, with the relaxation intervals $T_\parallel$ and T_1 for the longitudinal and transverse components respectively. Considering only small deviations of the magnetization M from the state M_0 in the inversely directed field

$$H_{0\Sigma} = \left(-H_0 - N_3 H_0 + \frac{2K_1}{\mu_0 M_0}\right)e_3,$$

we obtain the following expression for the dissipation term:

$$R = \frac{1}{T_1}\left(\frac{M_0}{|H_{0\Sigma}|}H_{\Sigma 1} + M_1\right)e_1 + \frac{1}{T_1}\left(\frac{M_0 H_{\Sigma 2}}{|H_{0\Sigma}|} + M_2\right)e_2 +$$
$$+ \frac{1}{T_\parallel}\left(\frac{M_0}{|H_{0\Sigma}|}H_{\Sigma 3} - M_3\right)e_3, \qquad (6)$$

where

$$|H_{0\Sigma}| = H_0(1 + \chi_0 N_\parallel), \quad \chi_0 = \frac{\mu_0}{H_0}, \quad N_\parallel = N_3 - \frac{2K_1}{\mu_0 M_0}.$$

This differs from the well-known Bloch-Wangsness expression (Mikhaelyan, 1963) in the signs of M_1 and M_2, which accounts for the reverse direction of the transverse relaxation process when $M_0 \cdot H_{0\Sigma} < 0$.

Let us consider a strictly longitudinal relaxation process, for which we shall assume in (1)–(6) the following:

$$H_1 = H_2 = M_1 = M_2 = 0, \ H_3 = -H_0, \ (H_0 > 0), \ M_3 = M_0(1 + m_3^0). \quad (7)$$

Substituting (7) into (1)–(6), we obtain, after linearization, the equation

$$T_\| \frac{dm_3^0}{dt} + \left(1 + \frac{\chi_0 N_0}{1 + \chi_0 N_\|}\right) m_3^0 = -2, \quad (8)$$

where $N_0 = N_3 + (2K_1/\mu_0 M_0)$. It follows that the magnetization varies almost linearly during the small period after the field is inverted:

$$m_3^0 \approx -2\frac{t}{T_\|}, \ M_3 = M_3^0 \approx M_0\left(1 - \frac{2t}{T_\|}\right), \ t \ll T_\|. \quad (9)$$

Now let us consider small perturbations imposed on the longitudinal relaxation process by setting

$$H_1 = H_0 h_1, \ H_2 = H_0 h_2, \ H_3 = -H_0(1 - h_3),$$
$$M_1 = M_0 m_1, \ M_2 = M_0 m_2, \ M_3 = M_0(1 + m_3^0 + m_3),$$
$$H_0 > 0, \ M_0 > 0, \ h_1, h_2, h_3 \ll 1, \ m_1, m_2, m_3 \ll 1,$$

where m_3^0 is found from (8) and (9). Using (1)–(6), (8) and (9), we arrive at the system of equations

$$(p - a_1)m_1 - b_2 m_2 = f_1 h_1 + g_0 h_2,$$
$$b_1 m_1 + (p - g_2)m_2 = -g_0 h_1 + f_1 h_2,$$
$$(p + c_0)m_3 = f_\| h_3, \quad (10)$$

where the following notation has been introduced:

$$p = \frac{d}{d\tau}, \ a_2 = \frac{1}{\tau_1}\left(1 - \frac{\chi_0 N_0}{1 + \chi_0 N_\|}\right),$$
$$b_i = 1 + \chi_0(N_\| - N_i) + \chi_0(N_0 - N_i)m_3^0, \ i = 1, 2;$$
$$c_0 = \frac{1}{\tau_\|}\left[1 + \frac{\chi_0 N_0}{1 + \chi_0 N_\|}(1 - 2m_2^0)\right], \ f_1 = \frac{1}{\tau_1(1 + \chi_0 N_\|)},$$
$$f_\| = \frac{1}{\tau_\|(1 + \chi_0 N_\|)}), \ g_0 = 1 + m_3^0, \ \tau_1 = \omega_H T_1, \ \tau_\| = \omega_H T_\|.$$

The solution of the system (10) can be expressed in the form $m = \hat{\chi}^0 h$, where $m = (m_1, m_2, m_3)$, $h = (h_1, h_2, h_3)$,

$$
\hat{\chi}^0 = \begin{bmatrix} \chi_{11}^0 & \chi_{12}^0 & 0 \\ \chi_{21}^0 & \chi_{22}^0 & 0 \\ 0 & 0 & \chi_{33}^0 \end{bmatrix} ;
$$

$\hat{\chi} = \chi_0 \hat{\chi}^0$, is the magnetic susceptibility tensor and

$$
\chi_{11}^0 = \frac{f_1(p - a_2) - g_0 b_2}{(p - a_1)(p - a_2) + b_1 b_2}, \quad \chi_{22}^0 = \frac{f_1(p - a_1) - g_0 b_1}{(p - a_1)(p - a_2) + b_1 b_2},
$$

$$
\chi_{33}^0 = \frac{f_{11}}{p + c_0},
$$

$$
\chi_{12}^0 = \frac{g_0(p - a_2) + f_1 b_2}{(p - a_1)(p - a_2) + b_1 b_2}, \quad \chi_{21}^0 = -\frac{g_0(p - a_1) + f_1 b_1}{(p - a_1)(p - a_2) + b_1 b_2}.
$$

An analysis of these expressions for the magnetic susceptibility tensor allows us to make the following conclusions:

(a) the state of inverse magnetization is unstable because the roots of the system's characteristic equation $(p - a_1)(p - a_2) + b_1 b_2 = 0$ possess a positive real part

$$
\mathrm{Re}\, P_\Sigma = \frac{a_1 + a_2}{2} = \frac{1}{\tau}\left(1 - \frac{N_1 + N_2}{2}\frac{\chi_0}{1 + \chi_0 N_\parallel}\right) > 0,
$$

since $N_\parallel > 0$, $(N_1 + N_2)/2 = N_1 < 1$; the instability is exhibited in the growth of the precession amplitude with an increment of the order of the reciprocal $1/T_\perp$ of the time of transverse relaxation;

(b) when $T_\perp$ is great enough, the transverse components of the static magnetic susceptibility ($P = 0$) have the negative values

$$
\chi_{11} = -\chi_0 \frac{g_0}{b_i} = -\chi_0 \frac{1 + m_3^0}{1 + \chi_0(N_\parallel - N_1) + \chi(N_0 - N_1)m_3^0} < 0, \qquad (11)
$$

which are greater than unity in absolute value (superdiamagnetism) when

$$
N_1 - N_\parallel > \frac{H_0}{M_0} - 1, \quad m_3^0 = 0.
$$

Since $N_i \leq 1$ and $N_\parallel > 0$, the last inequality can hold only when $H_0 < 2M_0$, and in this sense it is advantageous to experiment with the more elongated specimens when $H_\parallel$ is small. If

$$
\chi \frac{4K_1}{\mu_0 M_0} = 1,
$$

then in linear approximation the transverse susceptibility becomes independent
of m_3^0, and therefore remains constant during a small period after remagnetization. Consequently, a rapid change in the strength and sign of the magnetization field can result in a metastable superdiamagnetic state. Its stability
depends on the interaction between the magnetic and the environment, as well
as on the law of space-time distribution of the controlled process (Samoilenko,
1968, 1972).

6.3. Superdiamagnetism and Parametrically Stimulated Anomalous Gyrotropy

Kuts and Melnik (1974) and Kuts et al. (1973) showed that under the influence of Larmor frequency harmonic modulation with period comparable to the
relaxation time, states periodically occur in which transverse magnetization is
oriented inversely to the transverse field. It is of interest to find the conditions
under which the average magnetic permeability becomes negative. This can be
accomplished by analyzing the appropriate expression for the transverse susceptibility in longitudinal pumping. However, considering the slow convergence of
the series in terms of Bessel functions of increasing orders, it is easier to utilize
the technique of harmonic balance which gives a solution of the equation with
a periodic parameter in the form of a rapidly converging continuous fraction.

It is shown below that longitudinal parametrical pumping with a frequency
slightly higher than the eigenfrequency of magnetic resonance, provided the
modulation of the longitudinal field is deep enough, transfers the gyromagnetic
medium into a state with anomalous properties, namely, a negative averaged
static permeability and non-zero static gyrotropy. By the latter one means the
occurrence of a non-zero antisymmetric component of the susceptibility tensor
with respect to the constant transverse field. This can be advantageous, for
instance, for the stabilization of gyrosystems by means of non-potential forces
of radial correction.

In order to eliminate awkward calculations and gain a better insight into
the problem, we are going to examine only one of the simplest magnetostatic
modes, namely, the homogeneous precession realized by homogeneous magnetization of a specimen, although similar phenomena can occur when a specimen
is magnetostatically stimulated in some other fashion. It is assumed that the

gyromagnetic medium is a monocrystal specimen of ferrite oriented, as in the preceding case, along the unit vectors e_i and magnetized to saturation in the direction e_3. The modulation of the magnetizing field is such that the state of saturation is not disturbed when the field strength is at its minimum. Both the perturbation and the deviation from the equilibrium are assumed to be small enough.

The Cartesian components of the vectors M and H can be presented in the form

$$M_i = M_0 m_i, \quad H_i = H_0 h_i \ (i = 1, 2), \quad H_3 = H_0(1 - g \sin \Omega t), \qquad (1)$$

where $(m_i, h_i, g) \ll 1$. Given these conditions and a small enough damping parameter $0 < \alpha \ll 1$, we can use equation (2.1) with the dissipation term in the Landau-Lifshits form

$$R = -\frac{\alpha \gamma}{M} M \times (M \times H_\Sigma). \qquad (2)$$

Substituting (1) and (2) into (2.1)–(2.5), we obtain in linear approximation:

$$\begin{aligned}
\frac{dm_1}{d\tau} + \alpha(s_1 - q \sin \nu_0 \tau)m_1 + (s_2 - q \sin \nu_0 \tau)m_2 &= \alpha h_1 + p_2, \\
\frac{dm_2}{d\tau} - (s_1 - q \sin \nu_0 \tau)m_1 + \alpha(s_2 - q \sin \nu_0 \tau)m_2 &= -h_1 + \alpha h_2,
\end{aligned} \qquad (3)$$

where

$$s_1 = 1 - \chi(N_\parallel - N_1), \quad s_2 = 1 - \chi_0(N_\parallel - N_2),$$
$$N_\parallel = N_3 - \left(\frac{2k_1}{\mu_0 M_0}\right), \quad \tau = \omega_H t, \quad \omega_H = \gamma H_0, \quad \nu_0 = \frac{\Omega}{\omega_H}, \quad \chi_0 = \frac{M_0}{H_0}.$$

In the case of uniaxial shape anisotropy, when $N_1 = N_2 = N_\perp$, $s_1 = s_2 = s$, we can introduce complex variables $m = m_1 + i m_2$ and $h = h_1 + i h_2$; the system (3) is then reduced to a linear equation of the first order

$$\frac{dm}{d\tau} + (\alpha - i)(s - q \sin \nu_0 \tau)m = (\alpha - i)h. \qquad (4)$$

We note that when the right hand side of (4) is a harmonic function of time, $h = h_0 \exp(i\nu\tau)$, we can substitute $m = m_0 \exp(i\nu\tau)$ and arrive at an equation of the form (4) with a constant right hand side and a new constant parameter $s_0 = s - (\nu/(1 + i\alpha))$. Therefore it is advisable first to carry out an analysis for the case when the right hand side is either constant or varying slowly enough. For the linearized model to be valid, h is chosen small enough so that the amplitude of the induced transverse oscillations of magnetization does not exceed the excitation threshold for additional resonances and there is no non-linear detuning of free precession.

According to Monosov (1970), the precession angle is limited by the value $\theta_* = \frac{1}{2}(\Delta H/M_0)^{1/2}$ where ΔH is the half-width of the resonance range of the magnetizing field. The general integral of equation (4)

$$m(\tau) = m(0)\exp[-(\alpha - i)(s\tau + q\cos\nu_0\tau)]+$$
$$+(\alpha - i)\int_0^\tau \exp\{-(\alpha - i)[(\tau - \tau_1)s+$$
$$+q(\cos\nu_0\tau - \cos\nu_0\tau_1)]\}h(\tau_1)d\tau_1 \qquad (5)$$

shows that free motion in this case is asymptotically stable when $s > 0$. This condition holds if the ferrite is magnetized to saturation, that is, $H_0 > M_0 N_\| \times \times (1 - q)^{-1}$.

The properties of the slowly varying component of the induced motion in question do not clearly follow from the exact solution of (5) or the presentation of it as a double-row series of Bessel functions of q. These phenomena are clearer if the solution of (4) is found directly as an expansion including a quasiconstant component x and a number of harmonic multiples of ν_0 with slowly varying amplitudes x_n^+ and x_n^-:

$$m = x + \sum_{n=1}^{\infty}[x_n^+ \exp(in\nu_0\tau) + x_n^- \exp(-in\nu_0\tau)]. \qquad (6)$$

Substitution of (6) into (4) and the technique of harmonic balance (Bogolyubov and Mitropolskii, 1963) yield the recurrent system of equations

$$\blacksquare \qquad sx + \frac{q}{2i}(x_1^+ - x_1^-) = h,$$
$$\left(s \mp \frac{\nu_0}{1 + i\alpha}\right)x_1^\pm \mp \frac{q}{2i}(x - x_2^\pm) = 0,$$
$$\left(s \mp \frac{2\nu_0}{1 + i\alpha}\right)x_2^\pm \mp \frac{q}{2i}(x_1^\pm - x_3^\pm) = 0,\dots, \qquad (7)$$
$$\left(s \mp \frac{n\nu_0}{1 + i\alpha}\right)x_n^\pm \mp \frac{q}{2i}(x_{n-1}^\pm - x_{n+1}^\pm) = 0 \ (n \geq 2). \quad \square$$

Assuming $x_{n+1}^\pm = 0$, it is easy to solve this system in x expressing the result in terms of continuous fractions. Passing to the limit as $n \to \infty$, we obtain

$$\frac{x}{h} = \frac{1}{s}[1 - \epsilon(F^+ - F^-)]^{-1},$$

where

$$F^{\pm} = \left\{ \frac{\nu_0}{\beta s} \pm 1 - \epsilon \left[\frac{2\nu_0}{\beta s} \pm 1 - \epsilon \left(\frac{3\nu_0}{\beta s} \pm 1 - \ldots \right) \right]^{-1} \right\}^{-1}, \qquad (8)$$

$\beta = 1 + i\alpha$, $\epsilon = q^2/4s^2 \ll 1$ is a 'small' parameter characterizing the pumping power. Infinite continuous fractions $F^{\pm}$ converge rapidly if q is small enough, because their exact values are confined between the appropriate neighbouring fractions $F_n^{\pm}$ and $F_{n+1}^{\pm}$ which converge rapidly as $q \to 0$ and $n \to \infty$.

Let us introduce a dimensionless detuning $\xi = 1 - \nu_0 s^{-1}$. If we assume that ξ, α, and ϵ are of the same order, then to within ϵ, (8) yields:

$$\frac{x}{h} \simeq -\frac{1}{s} \left(\frac{\epsilon}{\xi + i\alpha} - 1 \right)^{-1} = -\frac{1}{s} \left[\frac{\xi(\epsilon - \xi) - \alpha^2}{(\epsilon - \xi)^2 + \alpha^2} + \frac{i\alpha\epsilon}{(\epsilon - \xi)^2 + \alpha^2} \right]. \qquad (9)$$

If the detuning $\xi = \epsilon - \alpha$, which can be regarded as optimal from the viewpoint of minimization of q, then

$$\frac{x}{h} = -\frac{1}{s} \left(\frac{\epsilon}{2\alpha} - 1 + i\frac{\epsilon}{2\alpha} \right). \qquad (10)$$

Reverting to the original variables of the problem and taking into account the well-known relation between the susceptibility and the permeability $\hat{\mu}_{\perp} = 1 + \hat{\chi}_{\perp}$ we obtain the final result from (10) in the form of the quasistatic transverse permeability tensor:

$$\hat{\mu}_{\perp} = \begin{vmatrix} -\mu_{(-)} & g \\ -g & -\mu_{(-)} \end{vmatrix}, \ \mu_{(-)} = \frac{\chi_0}{s} \left(\frac{\epsilon}{2\alpha} - 1 \right) - 1, \ g = \frac{\chi_0}{s} \frac{\epsilon}{2\alpha}. \qquad (11)$$

The property of parametrically stimulated gyromagnetic media formulated above stems from this directly. Indeed, suppose that $\epsilon > \epsilon_*$, where $\epsilon_* = 2\alpha(1 + s\chi_0^{-1})$ is the critical value of the pumping parameter characterizing the excitation threshold of the state in question. Now for the chosen detuning $\xi = \epsilon - \alpha$, the principal diagonal terms of the tensor $\mu_{\perp}$ become negative, which reveals the anomalous property with respect to constant or slowly varying perturbations. This property can be regarded as parametrically stimulated superdiamagnetism. Furthermore, when $\epsilon \neq 0$ and $\xi = \epsilon - \alpha$, another property can be observed: g, the parameter of static gyrotropy, differs essentially from zero.

Note that in view of (7) the quantity x_1^+ is of the same order as the quantity xg^{-1}, while the orders of x_n^+ ($n \geq 2$) are equal to xg^{n-2}. Therefore, the limitation of the precession angle imposes an overall limit on the amplitude

x_1^+ of the first harmonic. As was mentioned above, the greatest precession angle is approximately given by $\theta_* \approx \frac{1}{2}\sqrt{\Delta H/M_0}$. When $M_0 \sim H$, we have $\theta_* \sim \sqrt{\Delta H/H_0} = \sqrt{\alpha} \sim \sqrt{\epsilon} \sim q$. Hence the relative value of the constant component of transverse magnetization is limited by the value $x_* \sim q\theta_* \sim q^2 \sim \alpha$, that is, for the chosen detuning $\xi = \epsilon - \alpha$ the domain of linear negative permeability is comparatively small.

If we assume that $\alpha \ll \xi \sim 1$, then

$$\mu_{(-)} \approx \frac{\chi_0}{s}\left(\frac{\epsilon}{\xi} - 1\right) - 1.$$

Therefore $\mu_{(-)} > 0$ for $\epsilon > \epsilon_* = \xi(1 + s\chi_0^{-1})$. In consequence, the limit of linearity extends to a value of order $x_* \sim q\theta_* \sim \sqrt{\epsilon\alpha} \sim \sqrt{\alpha\xi} \sim \sqrt{\alpha}$.

Qualitatively, three basic processes are responsible for this phenomenon:

(1) The emergence of a periodic pumping component that is transverse to the resulting constant field and is due to the change in its direction when the transverse constant perturbing field is imposed.

(2) The resonance stimulation of the magnetic moment precession with a phase shift with respect to the pumping field, the shift being dependent on the chosen detuning.

(3) The synchronous interaction between the transverse oscillations of the magnetization and the longitudinal component of pumping, owing to which the precession axis inclines and the transverse component of the constant magnetization emerges. The direction of the vector of averaged (constant) magnetization in the transverse plane depends, in its turn, on the phase shift between the oscillations of magnetization and the pumping field. When the pumping frequency is below the resonance frequency, the transverse magnetization, averaged over the period of pumping, has a component that is opposite in direction to the imposed transverse field, and also has a component that is perpendicular to the transverse field.

Let us analyze the transition process of magnetization of the parametrically stimulated gyromagnetic medium in a superdiamagnetic state. Assume x and $x_n^{\pm}$ in (6) to be slowly varying functions of time. Then setting $p = d/dt$, we obtain, instead of (7), a system of equations where instead of s, we have the function $s_1 = s + ip/\beta$. For $p = i\nu$ we obtain the function s_0 introduced above, which follows naturally from the rules of operation calculus.

In order to be able to compare the orders of quantities with the small parameter α, we introduce the following notation: $\xi = \alpha\xi_0$, $\epsilon = \alpha\epsilon_0$, $p = \alpha s\tilde{p}$. Replacing s by s_1 in (9) and using the new notation, we have in first approximation in α the following expressions for the transfer functions:

$$\mathrm{Re}\left(\frac{x}{h}\right) = -\frac{1}{s}\left(\frac{\epsilon_0\xi_0}{\tilde{p}^2 + 2\tilde{p} + 1 + \xi_0^2} - 1\right), \tag{12}$$

$$\mathrm{Im}\left(\frac{x}{h}\right) = -\frac{\epsilon_0}{s}\frac{\tilde{p}+1}{\tilde{p}^2 + 2\tilde{p} + 1 + \xi_0^2}. \tag{13}$$

Now we can easily find the transition processes of magnetization by applying the inverse Laplace transform to (12) and (13) (Lappo–Danilevsky, 1957). For instance, for $\xi_0 = 1$

$$\mathrm{Re}\left(\frac{x}{h}\right)(\tau) = \frac{1}{s}\left[1 - \frac{\epsilon_0}{2}(1 - e^{-\alpha s\tau}\cos\alpha s\tau)\right].$$

Thus, if in the direction e_1 the field 'jumps' by H_1, the magnetization M_1 will vary with time as

$$M_1(t) = H_1\frac{\chi_0}{s}\left[1 - \frac{\epsilon}{2\alpha}(1 - e^{-\alpha s\omega_H t}\cos\alpha s\omega_H t)\right]. \tag{14}$$

At first, the directions of the field and the magnetization M_1 are the same, but after the transition process for $\epsilon > 2\alpha$ the direction of M_1 reverses, which corresponds to negative static susceptibility. The transition process time is of the order of the relaxation time, but it can be diminished by increasing ξ.

6.4. Low-Frequency Susceptibility of a Gyromagnetic Medium

The susceptibility of modulated gyromagnetic media with respect to a high-frequency field of transverse polarization was discussed in Kuts and Melnik (1974), Kuts et al. (1973), and in Fabrikov (1961). We are now going to analyze a quasiclassical model of the behaviour of low-frequency transverse susceptibility of magnetic media when a high-frequency longitudinal pumping is applied. If the frequency of pumping is close to the eigenfrequency of the magnetic resonance, the medium acquires special properties with respect to the low-frequency signal: either positive or negative dissipation can be obtained depending on the sign of the detuning ξ defined as the difference between the pumping frequency ω_H and the resonance frequency ω_0 of the specimen magnetized to saturation. The requirement for negative dissipation is $\omega_H > \omega_0$.

Static susceptibility can become negative in this case as well and can even be greater than one in absolute value, which can be regarded as a 'superdiamagnetic' state.

Unlike what is covered in Fabrikov (1958), this can be accounted for by the synchronous nutation phenomenon. The term 'synchronous' implies here that the frequency of the Larmor precession of the magnetization vector and the frequency of the modulating oscillations are the same. For the sake of convenience, let us take an isotropic ferromagnetic specimen magnetized to saturation; if the specimen is ellipsoidal, it simplifies the problem but does not make the results less general. Consider the canonical equation of homogeneous precession of the magnetic moment b_0 of the ferromagnetic (Lvov and Strabinets, 1971):

$$\dot{b}_0 + \eta_0 b_0 = i\frac{\delta \mathrm{H}}{\delta b_0^*}, \tag{1}$$

where H is the Hamiltonian of the medium. The term $\eta_0 b_0$ was introduced into this equation phenomenologically in order to take account of the relaxation $\eta_0 = \gamma \Delta H_0/2$ where ΔH_0 is the width of the resonance curve and γ is the magnetomechanical ratio. Using the inverse Holstein-Primakov transformation, it is easy to find from (1) the equation of motion for the density of the magnetic moment. Taking the Zeeman and dipole energies into account, we arrive at a Hamiltonian of the following form:

$$\mathrm{H} = \omega_0 b_0 b_0^* + \mathrm{H}_p(t) + V_s, \tag{2}$$

where H_p represents the interaction between the homogeneous precession c and the external energy source, that is, the pumping field $h(t)$; V_s allows for the influence of the transverse magnetic field of the signal h_s on the homogeneous precession. Analyzing the case of the space-homogeneous magnetic pumping, let us make use of the well-known (Schlömann, 1964) expression for the Hamiltonian

$$\mathrm{H}_p = \frac{1}{2}\left[\gamma h(t)\left(\frac{A_0}{\omega_0}b_0 b_0^* - \frac{B_0}{\omega_0}b_0^* b_0^*\right) + \text{c.t.}\right], \tag{3}$$

where c.t. stands for the 'conjugate term',

$$A_0 = \gamma_0 H_0 - \omega_H N_z + (\omega_H/2)(N_x + N_y), \quad B_0 = (\omega_H/2)(N_x - N_y),$$
$$\omega_0 = (A_0^2 - B_0^2)^{1/2}, \quad \omega_H = 4\pi\nu M,$$

H_0 is a constant magnetizing field directed along the z-axis; N_x, N_y, N_z are the demagnetizing factors and M is the saturation magnetization. In the system of spherical polar coordinates, where the z-axis is directed along the magnetization vector M, the transverse demagnetizing factors N_x and N_y are equal. This leads to the following: $B_0 = 0$ and $A_0 = \omega_0$. Formula (3) now takes the form

$$\mathbf{H}_p = \frac{1}{2}[\gamma h(t)bb_0^* + \text{c.t.}]. \tag{4}$$

If the perturbation field is transverse, then

$$V_s = U h_s b_0^* + \text{c.t.} \tag{5}$$

Here $h_s = h_x + ih_y$ and U is the coupling factor.

The periodical time-dependent part of the Hamiltonian defines the parametrically stimulated state of the medium. The spectrum of the system's eigenfrequencies acquires side frequencies that differ from the principal one by the pumping frequency. Therefore, the solution of equation (1) can be expanded in a Fourier series. Substituting the explicit expression of the Hamiltonian into (1) and introducing the small perturbation parameter α, we obtain the dynamical equation in its linear approximation in b_0:

$$\dot{b}_0 + \alpha^2 \eta_0 b_0 = i[\omega_0 b_0 + \alpha \gamma h(t) b_0 - U h_s]. \tag{6}$$

Suppose that the pumping field $h(t) = h\cos(\omega_H t)$, whose frequency is $\omega_H = \omega_0 + \alpha^2 \xi$, is applied parallel to H_0, and that there is a low-frequency field of the perturbation signal $h_s = h_0 \exp(i\alpha^2 \Omega t)$. The solution of equation (6) can be found in the form

$$b_0 = \sum_n c_n \exp[i(n\omega_H + \alpha^2 \Omega)t], \quad n = 0, 1, 2, \ldots \tag{7}$$

Substituting (7) into (6) and taking into account the expressions for $h(t), h_s$ and ω_0, we obtain the recurrent system of equations

$$[(n-1)\omega_H + \alpha^2(\Omega + \xi - i\eta_0)]c_n - \frac{\alpha \gamma h}{2}(c_{n+1} + c_{n-1}) + U\delta_{n,0}h_0 = 0. \tag{8}$$

We use the technique of harmonic balance (Bogolyubov and Mitropolskii, 1963) in order to solve the system (8). To accommodate the conditions of the balance in α we set the amplitude of the resonance harmonic as $c_1 = \alpha^{-1}Y_1$.

Then the first approximation in α yields a system of three equations for the three harmonics $n = 0, 1, 2$. The amplitudes of the other harmonics in this approximation are assumed to be zero. This system allows us to find

$$-c_0 = \frac{\gamma h Y_1}{2} + U h_0, \tag{9}$$

where

$$Y_1 = \frac{\gamma h}{2} \frac{1}{\omega_H} U h_0 \frac{1}{\Omega + \xi - i\eta_0}.$$

Equation (9) yields the transfer function

$$W(p) = 1 - \left(\frac{\gamma h}{2}\right)^2 \frac{1}{\omega_H} \frac{1}{p + \eta_0 + i\xi}, \tag{10}$$

where $p = i\Omega$ is a complex parameter. Using the transfer function $W(p)$, we can easily find the components of the magnetic susceptibility tensor with respect to the low-frequency field of external perturbation:

$$\begin{bmatrix} m_x \\ m_y \end{bmatrix} = \chi_0 \begin{bmatrix} \operatorname{Re} W(p) & \operatorname{Im} W(p) \\ -\operatorname{Im} W(p) & \operatorname{Re} W(p) \end{bmatrix} = \begin{bmatrix} \chi_\perp & \chi_\alpha \\ -\chi_\alpha & \chi_\perp \end{bmatrix} \begin{bmatrix} h_x \\ h_y \end{bmatrix}. \tag{11}$$

Here

$$\chi_\perp(p) = \chi_0 \left[1 - \left(\frac{\gamma h}{2}\right)^2 \frac{1}{\omega_H} \frac{\xi}{p^2 + 2\eta_0 p + \eta_0^2 + \xi^2} \right], \tag{12}$$

$$\chi_\alpha(p) = -\chi_0 \left(\frac{\gamma h}{2}\right)^2 \frac{1}{\omega_H} \frac{p + \eta_0}{p^2 + 2\eta_0 p + \eta_0^2 + \xi^2}, \tag{13}$$

where $\chi_0 = U M_0 / \omega_H$.

It has to be pointed out that the expressions (12) and (13) were obtained using the asymptotic technique and therefore they give a satisfactory approximation within the accuracy α^2 in the region of low frequencies only. Although better approximations in α improve the accuracy of the result, the error still increases with a gain in the perturbation signal frequency. Inasmuch as we have confined ourselves to a study of the quasistatic case ($\Omega \ll \omega_0$), the discussion of the results should take into account the possible influence of the remaining terms on the calculation error in the high-frequency domain.

Let us write down separately the real and imaginary parts (that is, the dispersion and the dissipation parts respectively) of the transverse component $\chi_\perp = \chi'_\perp - i\chi''_\perp$:

$$\chi'_\perp = \chi_0 \left[1 - \left(\frac{\gamma h}{2}\right)^2 \frac{1}{\omega_H} \frac{\xi(\eta_0^2 + \xi^2 - \Omega^2)}{(\eta_0^2 + \xi^2 - \Omega^2)^2 + 4\Omega^2\eta_0^2} \right] + o(\alpha^2, \Omega), \tag{14}$$

$$\chi_\perp'' = -\chi_0 \left(\frac{\gamma h}{2}\right)^2 \frac{1}{\omega_H} \frac{2\xi\Omega\eta_0}{(\eta_0^2 + \xi^2 - \Omega^2)^2 + 4\Omega^2\eta_0^2} + o(\alpha^2, \Omega). \tag{15}$$

An essential property of these expressions is their resonance dependence on the frequency Ω of the external perturbation signal alternating field. The behaviour of the transverse component $\chi_\perp$ is similar to the behaviour of the diagonal component of the behaviour of the high-frequency magnetic susceptibility in ferromagnetic resonance, but in the present instance the resonance frequency ω_0 is less than the usual value by the pumping frequency ω_H. The value of $\chi_\perp$ is also essentially related to the detuning ξ. It follows from (15) that, depending on the sign of ξ, the dissipation part of the transverse component of the susceptibility tensor can be either positive or negative. In the case of $\xi > 0$, we can have maximum susceptibility if the resonance condition $\Omega^2 = \eta_0^2 + \xi^2$ holds in (15):

$$\chi_{\perp\max} = -\chi_0 \left(\frac{\gamma h}{2}\right)^2 \frac{1}{\omega_H} \frac{1}{2\eta_0} + o(\alpha^2, \Omega). \tag{16}$$

We can also find the negative static susceptibility from (14):

$$\chi_{\perp CT} = \chi_0 \left[1 - \left(\frac{\gamma h}{2}\right)^2 \frac{1}{\omega_H} \frac{1}{2\eta_0}\right] + o(\alpha^2, \Omega). \tag{17}$$

Thus the analysis of the quasiclassical model (1) has revealed that as in the classical case (§6.3), longitudinal parametrical pumping and a deep enough modulation of the longitudinal field make the ferromagnetic crystal transfer to a state with anomalous properties. In the low-frequency domain, there occur negative dissipation and anomalous dispersion of the transverse susceptibility. We see that in principle it is possible to obtain negative values of static susceptibility greater than unity in absolute value. This is a state of increased reactive power. The crystal behaves like a diamagnetic with increased negative susceptibility.

6.5. Stability of Spin Waves in Longitudinal Pumping of Ferromagnetic Crystals

The negative magnetic susceptibility in quasistatic or low-frequency magnetization is a matter of interest not only as an anomalous state of a magnetic but also as a useful property of the continuous medium that enables one to carry out field amplification with antiphase reflection. However, the negative magnetic susceptibility can only be achieved in the absence of parametrically unstable homogeneous precession with respect to the excitation of a pair of spin waves (Morgenthaller, 1960; Osborne, 1945; Patton, 1969). This is why it is necessary to investigate the stability of spin waves. We should also find out the relation between the susceptibility and the wave vector.

It is shown in this section that elimination of spin wave instability is possible if the pumping frequency is the sum of two harmonics. This will bring us to the negative susceptibility as a function of the wave vector in the low-frequency domain. Theoretical aspects were developed here for ellipsoidal specimens magnetized to saturation along one of the principal axes (Samoilenko and Khorozov, 1977).

We derive the equation of motion for the amplitudes of spin waves in the form that is somewhat different from the preceding one, which is more advantageous for the analysis of stability. Consider the equation of motion for the magnetization vector M in terms of the unit vector $m = M/M_0$:

$$\frac{d}{dt} m = -\gamma m \times H_\Sigma + R, \tag{1}$$

where γ is the magnetomechanical ratio, H_Σ is the total internal magnetic field and R is the dissipation term. The internal field consists of a permanent external field H_0, an external microwave field h_0, an effective exchange field H_{ex}, an effective field of dipole interaction H_d and an external perturbation field h. The dissipation term in equation (1) can be written as

$$R = \frac{\alpha^2}{T_2}(m_x l_x + m_y l_y) - \frac{\alpha^2}{T_1}(m_z - 1)l_z, \tag{2}$$

where $1/T_2 = \gamma \Delta H$. We expand the vector $m(r,t)$ in a spatial Fourier series whose coefficients are functions of time:

$$m(r,t) = \sum_k m_k(t)e^{ikr}, \tag{3}$$

where k is the spin wave vector.

Assuming that H_0 and h_0 are homogeneous within the volume of the specimen and directed along the z-axis and that $M_z = M_0$, we arrive at the equation of motion for m_k. According to Schlöman et al. (1960), the external field can be expressed in terms of the internal fields:

$$H = H_0 - 4\pi M N_z, \tag{4}$$

where M_z is the demagnetizing factor of the specimen. The result can be expressed in terms of the complex variables

$$a_k = m_k^x + i m_k^y, \quad a_{-k}^* = m_k^x - i m_k^y, \tag{5a}$$

$$k_1 = k_x + i k_y, \tag{5b}$$

$$h = h_x + i h_y. \tag{5c}$$

Considering the well-known low-power microwave experiments, we see that only the homogeneous mode of large amplitude can be excited while a_k cannot, apart from $k = 0$. Therefore we shall only have to deal with the terms of the first order in a_k, where $k \neq 0$. The equation of motion for a_k ($k \neq 0$) has the following form (Akhiezer et al., 1970; Monosov, 1970):

$$\dot{a}_k = i\{(A_k + C_k)a_k + [B_k e^{2i\Phi_k} + D_k]a_{-k}^*\} - \alpha^2 \gamma \Delta H_k a_k + i\alpha \gamma h_0^z a_k - i\gamma h_k, \tag{6}$$

where

$$A_k = \gamma H + \gamma D k^2 + \frac{1}{2}\omega_{\mathrm{H}} \sin^2 \theta_k, \tag{7a}$$

$$B_k = \frac{1}{2}\omega_{\mathrm{M}} \sin^2 \theta_k, \tag{7b}$$

$$C_k = -\frac{1}{2}\omega_{\mathrm{M}} \cos \theta_k \sin \theta_k [e^{-i\Phi_k} a_0 + e^{i\Phi_k} a_0^*] + \\ + \frac{\omega_{\mathrm{M}}}{4}[2N_z + (3\cos^2 \theta_k - 1)]a_0^2, \tag{7c}$$

$$D_k = \frac{1}{2}\gamma D k^2 a_0^2 - \omega_{\mathrm{M}} \cos \theta_k \sin \theta_k e^{i\Phi_k} a_0 + \\ + \frac{1}{4}\omega_{\mathrm{M}}(2\cos^2 \theta_k a_0^2 - \sin^2 \theta_k e^{2i\Phi_k} a_0^2), \tag{7d}$$

$$\omega_{\mathrm{M}} = \gamma 4\pi M, \quad \sin \theta_k = \frac{k_\perp}{k}, \quad \tan\Phi_k = \frac{k_x}{k_y},$$

D is a phenomenological exchange parameter defined by $H_{\mathrm{ex}} = D\nabla^2 m$, h_k is the amplitude of the perturbation field and α is a small perturbation parameter.

Inasmuch as the parametrical excitation of spin waves only brings about a small deviation of the magnetic moment from its equilibrium, equation (6) can be rewritten:

$$\dot{a}_k = i\{A_k a_k + B_k e^{2i\Phi_k} a_k^*\} - \alpha^2 \gamma \Delta H_k a_k + i\alpha \gamma h_0^z a_k - i\gamma h_k. \tag{8}$$

Now we can use the linear canonical Holstein-Primakov transformation from the circular variables a_k to the elliptical ones b_k:

$$b_k = \lambda_k a_k + \mu_k a_{-k}^*, \ \ \lambda_k = \cosh\left(\frac{\psi_k}{2}\right), \ \ \mu_k = \sinh\left(\frac{\psi_k}{2}\right)e^{2i\Phi_k}, \qquad (9)$$

where $\cosh\psi_k = A_k/\omega_k$, $\sinh\psi_k = B_k/\omega_k$ and

$$\omega_k = (A_k^2 - B_k^2)^{1/2} = [(\gamma H + \gamma Dk^2)(\gamma H + \gamma Dk^2 + \omega_M \sin^2 \theta_k)]^{1/2}. \qquad (10)$$

This transformation is possible if $|A_k| > |B_k|$, that is, $H > 4\pi N_z M$. If this is not the case, the ferromagnetic is unstable with respect to the appearance of the domain structure.

The equation of motion in b_k takes the form

$$\dot{b}_k = i\left\{\omega_k b_k + \alpha\gamma h_0^z(t)\frac{A_k}{\omega_k}b_k - \alpha\gamma h_0^z(t)\frac{B_k}{\omega_k}e^{2i\Phi_k}b_{-k}^*\right\} - \alpha^2\gamma\Delta H_k b_k - i\Omega h_k. \ (11)$$

Now let us analyze the stability of the spin waves. A necessary and sufficient condition is that the roots of the characteristic equation in the plane of the complex parameter $p = -i\omega$ be in the left half-plane. Let us find the solution of the equation system consisting of equation (10) and its conjugate equation

$$\dot{b}_k + \alpha^2\gamma\Delta H_k b_k = i\left\{\omega_k b_k + \alpha\gamma h_0^z(t)\frac{A_k}{\omega_k}b_k - \alpha h_0^z(t)\frac{B_k}{\omega_k}e^{2i\Phi_k}b_{-k}^* - i\Omega h_k\right\}, \tag{12}$$

$$\dot{b}_{-k}^* + \alpha^2\gamma\Delta H_k b_{-k}^* = i\left\{-\omega_k b_{-k}^* + \alpha\gamma h_0^z(t)\frac{B_k}{\omega_k}e^{-2i\Phi_k}b_k - \right.$$
$$\left. -\alpha\gamma h_0^z(t)\frac{A_k}{\omega_k}b_{-k}^* + i\Omega h_k^*\right\}.$$

The external microwave field and the perturbation field are respectively given by

$$h_0^z(t) = h_1 \cos \Omega t + h_3 \cos 3\Omega t, \qquad (13)$$

$$h_k = h_0(\omega, k)e^{-i(\alpha^2\omega t - kr)}. \qquad (14)$$

Suppose that the ferromagnetic crystal is pumped at the frequency Ω which is related to ω_k by the formula

$$\omega_k = \Omega + \alpha^2\xi_k, \qquad (15)$$

where ξ is the detuning. We look for a solution of the system (12) in the form

$$b_k = \sum_k b_n^0(\omega, k) \exp\{-i(n\Omega + \alpha^2\omega)t\},$$

$$b_{-k}^* = \sum_n b_{-n}^{0*}(-\omega, -k) \exp\{-i(n\Omega + \alpha^2\omega)t\}. \tag{16}$$

The expressions (13)–(16) are substituted into (12) to obtain

$$[(h+1)\Omega + \alpha^2(\omega + \xi_k + i\gamma\Delta H_k)]b_n^0 +$$

$$+\frac{A_k}{\omega_k}\left\{\frac{\alpha\gamma h_1}{2}(b_{n+1}^0 + b_{n-1}^0) + \frac{\alpha\gamma h_3}{2}(b_{n+3}^0 + b_{n-3}^0)\right\} -$$

$$-\frac{B_k}{\omega_k}e^{2i\Phi_k}\left\{\frac{\alpha\gamma h_1}{2}(b_{-n-1}^0 + b_{-n+1}^{0*}) + \frac{\alpha\gamma h_3}{2}(b_{-n-3}^{0*} + b_{-n+3}^{0*})\right\} -$$

$$-\Omega\delta_{n,0}h_0(\omega, k) = 0, \tag{17}$$

$$(h-1)\Omega + \alpha^2(\omega - \xi_k + i\gamma\Delta H_k)]b_{-n}^{0*} -$$

$$-\frac{A_k}{\omega_k}\left\{\frac{\alpha\gamma h_1}{2}(b_{-n-1}^{0*} + b_{-n+1}^{0*}) + \frac{\alpha\gamma h_3}{2}(b_{-n-3}^{0*} + b_{-n+3}^{0*})\right\} +$$

$$+\frac{B_k}{\omega_k}e^{-2i\Phi_k}\left\{\frac{\alpha\gamma h_1}{2}(b_{n+1}^0 + b_{n-1}^0) + \frac{\alpha\gamma h_3}{2}(b_{n+3}^0 + b_{n-3}^0)\right\} +$$

$$+\Omega\delta_{n,0}h_0^*(-\omega, -k) = 0 \tag{18}$$

We introduce the notation for the amplitudes of the resonance harmonics: $\alpha b_{-1}^0 = Y_{-1}, \alpha b_{-1}^{0*} = \overline{Y}_1$. Substituting $\bar{n} = 0, \pm1, \pm2, \pm3, \pm4$ into equations (17) and (18), reducing by α, and setting $\alpha = 0$ in every equation, we obtain the zero approximation system of equations in α which is solvable with respect to Y_{-1} and Y_1:

$$\blacksquare \quad \left\{\omega + i\gamma\Delta H_k + \xi_k - \frac{B_k}{\omega_k^2\Omega}\gamma^2\left(\frac{h_1^2}{3} - \frac{h_3^2}{5}\right)\right\}Y_{-1} +$$

$$+\frac{A_k B_k}{\omega_k^2\Omega}e^{2i\Phi_k}\gamma^2\left(\frac{h_1^2}{2} - \frac{h_1 h_3}{3}\right)Y_1 = -\gamma\frac{h_1}{2}\left(\frac{A_k}{\omega_k}h_0 - \frac{B_k}{\omega_k}e^{2i\Phi_k}h_0^*\right),$$

$$\left\{\omega + i\gamma\Delta H_k - \xi_k + \frac{B_k^2}{\omega_k^2\Omega}\gamma^2\left(\frac{h_1^2}{3} - \frac{h_3^2}{5}\right)\right\}\overline{Y}_1 -$$

$$-\frac{A_k B_k}{\omega_k^2\Omega}e^{-2i\Phi_k}\gamma^2\left(\frac{h_1^2}{2} - \frac{h_1 h_3}{3}\right)Y_{-1} = -\gamma\frac{h_1}{2}\left(\frac{B_k}{\omega_k}e^{-2i\Phi_k}h_0 - \frac{A_k}{\omega_k}h_0^*\right). \,\square \tag{19}$$

Substituting $p = -i\omega$, we rewrite the characteristic equation of the system (19):

$$p^2 + 2(\gamma\Delta H_k)p + (\gamma\Delta H_k)^2 + \left[\xi_k - \frac{B_k}{\omega_k^2\Omega}\gamma^2\left(\frac{h_1^2}{3} - \frac{h_3^2}{5}\right)\right] -$$

$$-\left(\frac{A_k - B_k}{\omega^2\Omega}\right)^2\gamma^2\left(\frac{h_1^2}{2} - \frac{h_1 h_3}{3}\right)^2 = 0. \tag{20}$$

Since $\gamma \Delta H_k > 0$, the condition of stability is that the free term of the characteristic equation must be positive:

$$
\Delta(\theta_k, k) = (\gamma \Delta H_k)^2 + \left[\gamma \Delta H_0 - \frac{\gamma D_k^2}{\alpha^2} + \frac{B_k^2}{\omega_k^2 \Omega} \gamma \left(\frac{h_1^2}{3} - \frac{h_3^2}{5}\right)\right] -
$$
$$
- \left(\frac{A_k B_k}{\omega_k^2 \Omega}\right)^2 \gamma^2 \left(\frac{h_1^2}{2} - \frac{h_1 h_3}{3}\right)^2 > 0, \qquad (21)
$$

where, as will be shown below, $\xi_k = -\gamma \Delta H_0 + \gamma(D_k^2/\alpha^2)$.

Analyzing (21), we can see that the positive part is at its minimum when $k = k_1$, that is,

$$
\gamma \frac{D}{\alpha^2} k_1^2 = \gamma \Delta H_0 + \frac{B_k^2}{\omega_k^2 \Omega} \gamma^2 \left(\frac{h_1^2}{3} - \frac{h_3^2}{5}\right),
$$

while the negative part is at its maximum when $\theta_k = \pi/2$. Now if $h_3 = 3h_1/2$, we obtain $\Delta(\pi/2, k_1)_{\min} = (\gamma \Delta H_k)^2 > 0$. Therefore the system is stable.

Let us find the magnetic susceptibility of the medium with respect to the perturbation signal. Substituting $n = 0$ into equations (17) and (18), we obtain

$$
b_0^0(\omega, k) = -\frac{A_k}{\omega_k} \frac{\gamma h_1}{2\Omega} Y_{-1} + \frac{B_k}{\omega_k} e^{2i\Phi_k} \frac{\gamma h_1}{2\Omega} \overline{Y}_1 + h_0(\omega, k),
$$
$$
b_0^{*0}(-\omega, -k) = -\frac{A_k}{\omega_k} \frac{\gamma h_1}{2\Omega} \overline{Y}_1 + \frac{B_k}{\omega_k} e^{-2i\Phi_k} \frac{\gamma h_1}{2\Omega} Y_{-1} + h_0^*(-\omega, -k).
$$
$$(22)$$

For instance, let us consider the case $\theta = 0$. It follows from (7a) and (7b) that $A_k = \gamma H + \gamma D k^2$, $B_k = 0$, $\omega_k = A_k$; then, after solving the system (19), we obtain

$$
Y_{-1} = -\frac{\gamma h_1/2}{\omega + i\gamma \Delta H_k + \xi_k} h_0, \quad \overline{Y}_1 = \frac{\gamma h_1/2}{\omega + i\gamma \Delta H_k - \xi_k} h_0^*. \qquad (23)
$$

Substitution of (23) into (22) yields

$$
\frac{b_0^0}{h_0} = 1 + \frac{\gamma^2 h_1^2}{4\Omega} \frac{-i}{-i\omega + \gamma \Delta H_k - i\xi_k},
$$
$$
\frac{b_0^{0*}}{h_0} = 1 + \frac{\gamma^2 h_1^2}{4\Omega} \frac{i}{-i\omega + \gamma \Delta H_k + i\xi_k}.
$$
$$(24)$$

We define the transfer function as

$$
W(p, k) = 1 - \frac{\gamma^2 h_1^2}{4\Omega} \frac{i}{p + \gamma \Delta H_k - i\xi_k}, \qquad (25)
$$

where $p = -i\omega$. Returning to our original variables, that is, from the elliptical variables b_k back to the circular ones a_k, and recalling that $a_k = m_k^x + i m_k^y$, we can easily find the elements of the susceptibility tensor by expressing m_x and m_y in terms of h_x and h_y by means of the transfer function:

$$\chi'_\perp(p, k) = \chi_0 \frac{\gamma^2 h_1^2}{4\Omega} \frac{\xi_k}{p^2 + 2\gamma\Delta H_k p + \gamma^2 \Delta H_k^2 + \xi_k^2}, \tag{26}$$

$$\chi_a = \chi_0 \frac{\gamma^2 h_1^2}{4\Omega} \frac{p + \gamma\Delta H_k}{p^2 + 2\gamma\Delta H_k p + \gamma^2 \Delta H_k^2 + \xi_k^2}. \tag{27}$$

The static magnetic susceptibility has the form

$$\chi_\perp(0, k) = \frac{\gamma^2 h_1^2}{4\Omega} \frac{\chi_0 \xi_k}{\gamma^2 \Delta H_k^2 + \xi_k^2} + \chi_0. \tag{28}$$

It follows from equation (28) that the static susceptibility is negative when $\xi_k < 0$. It is at its maximum absolute value when $|\xi_k| = \gamma\Delta H_k$. We can assume $\xi_0 = -\gamma\Delta H_0$, therefore we obtain

$$\xi_k = -\gamma\Delta H_0 + \gamma\frac{Dk^2}{\alpha^2}. \tag{29}$$

Therefore

$$\chi'_\perp = -\chi_0 \frac{\gamma^2 h_1^2}{4\Omega} \frac{\gamma\Delta H_0 - \frac{\gamma Dk^2}{\alpha^2}}{p^2 + 2\gamma\Delta H_k p + \gamma^2 \Delta H_k^2 + \left(\gamma\Delta H_0 - \gamma\frac{Dk^2}{\alpha^2}\right)^2}, \tag{30}$$

$$\chi_a = \chi_0 \frac{\gamma^2 h_1^2}{4\Omega} \frac{p + \gamma\Delta H_k}{p^2 + 2\gamma\Delta H_k p + \gamma^2 \Delta H_k^2 + \left(\gamma\Delta H_0 - \gamma\frac{Dk^2}{\alpha^2}\right)^2}. \tag{31}$$

It follows that the magnetic susceptibility is negative only when $k^2 < \alpha^2 \Delta H_0/D$.

Considering the question of the reflection of the low-frequency perturbation signal from a specimen possessing a negative magnetic susceptibility, we can make use of the formula for the reflection coefficient (Gurevich, 1973)

$$\Gamma = \frac{\zeta_2^\pm - \zeta_1^\pm}{\zeta_2^\pm + \zeta_1^\pm}, \tag{32}$$

where $\zeta_{1,2}^\pm = \pm i(\mu_{1,2} \pm \mu_{a1,2})^{1/2}$. Since $\zeta_2^\pm < 1$, it follows from (32) that $|\Gamma| > 1$, that is, the reflected field is amplified with the reflection in the antiphase.

Thus we can make the following conclusions:

(1) When a ferromagnetic crystal is pumped longitudinally, and the pumping is the sum of the first and the third harmonics, the amplitudes of which have the ratio $h_1/h_3 = 2/3$, one can achieve absolute stability of the spin waves.

(2) The dependence of the susceptibility on the wave-vector is such that negative susceptibility can occur only when $k^2 < \alpha^2 \Delta H_0 / D$.

6.6. Applications

The above analysis of the methods of achieving the states with negative magnetic permeability has shown that the use of magnetic resonance for distributed control is possible in principle but there are certain difficulties. First of all, the precession angle is limited by the value θ_* and therefore, the dynamic range of amplification is limited. Besides, the method of synchronous parametrical pumping requires a large value for the reactive power to sustain the superdiamagnetic state. This power increases with a gain in the desired performance rate achieved due to an increase in the detuning ξ. The technique of inverting the magnetizing field yields a performance rate comparable to the period of precession, but there is a relatively fast loss of superdiamagnetism after remagnetization.

These difficulties can be partially overcome when the control media are applied under concrete conditions. In particular, θ_* can be increased by choosing special magnetics designed for operating in high-power microwave fields, for example, some of the nickel ferrochromites. It is also possible to employ impulses such that additional resonances do not noticeably show up. Special dynamical additional magnetization is also known to be used for the purpose.

The problems arising from the great level of the reactive power required for pumping can be overcome in some cases with the aid of superconducting resonators with a pulse energy take-off. A combination of the method of parametrical pumping and the method of field inversion can also prove to be efficient. In this case it is advisable to apply a system with the inverted field as the principal controller and a medium with parametrical pumping as the subsystem stabilizing the principal system. This makes it possible to use low levels of pumping power and, because of the stabilization, to prolong the superdiamagnetic state with reversed magnetization.

APPENDIX 1

Mathematical Models of Quantum Processes

This is a short presentation of the mathematical formalism and quantum-mechanical models used in the book. For simplicity's sake, we are not going to consider the relativistic problems. However, the general approach to control problems in the relativistic case is the same.

The theory of quantum mechanics is employed to describe the processes occurring in the microcosm and to evaluate their macroscopic manifestations. In principle, quantum mechanics allows us to find the current average values of physical quantities, the probabilities or probability amplitudes of individual events if there is a complete description of the system's state at the initial instant. The state of a quantum system with complete information (that is, with as much information as is physically possible) can be described by norm-preserving elements ψ of a complex Hilbert space, and the physical variables of the system are associated with self-adjoint (Hermitian) operators $\widehat{A}$ in the same space. (For brevity, operators are often identified with the physical quantities they represent.) It is noteworthy that in most cases of practical interest, a complete description of the system is provided by a complete set of simultaneously measurable physical quantities $\widehat{A}_i$, $i \in 1, \ldots, N$ specific to the system. Simultaneous measurement of these and other physical quantities does not give rise to uncertainty only when their operators commute with the operators of the above-mentioned complete set. Any consideration of operators that do not belong to the complete set of physical quantities (mechanical variables) requires establishing the rules of their commutation with the operators of the complete set.

173

The classical physical quantities that are directly observable in macroscopic measurements as the result of averaging over a large number of experiments or over the ensemble are associated with the quadratic forms

$$A = (\psi, \widehat{A}\psi), \tag{1}$$

where $(\cdot, \cdot)$ is the symbol of the scalar product of two elements in the complex Hilbert space. The evolution of the system can be observed on the macroscopic level only as a time variation of the averaged quantities (expected values) $\bar{A} = A(t)$.

The time variation of an averaged quantity can be described equivalently by means of various representations of the change of state. In the *Schrödinger representation*, the state is represented by a vector (element) ψ and varies under the action of the evolution operator $\widehat{S}(t)$ but the operators of the physical quantities do not evolve, although they can depend explicitly on time as a parameter, denoted by the subscript t:

$$\psi(t) = \widehat{S}(t)\psi(0), \quad \widehat{A}_t(t) = \widehat{A}_t(0) = \widehat{A}_t. \tag{2}$$

In the *Heisenberg representation*, the same evolution operator $\widehat{S}(t)$ carries out a similarity transformation of the operators of physical quantities (dynamic variables), which can also depend explicitly on time as a parameter, although the state vector does not vary with time, so that

$$\widehat{A}_t(t) = \widehat{S}^{-1}(t)A_t(0)\widehat{S}(t), \quad \psi(t) = \psi(0) = \psi. \tag{3}$$

In both of these cases, the evolution operator defines a continuous group of transformations of the Hilbert space into itself, the generating operator being the energy operator (Hamiltonian) of the considered system with the coefficient $-i/\hbar$, where $2\pi\hbar = h \simeq 6.627 \cdot 10^{-34} J \cdot s = 6.627 \cdot 10^{-27} \text{erg·s}$ is Planck's constant. Therefore, the evolution operator is defined by the equation

$$i\hbar \frac{d\widehat{S}(t)}{dt} = \widehat{H}_t \widehat{S}(t) \tag{4}$$

with the initial condition $\widehat{S}(0) = \hat{I}$, where $\hat{I}$ is the identity operator. In equation (4), the Hamiltonian $\widehat{H}_t = \widehat{H}_t(0)$ does not transform in the process of the system's evolution, although its explicit dependence on time, if any, must be taken into account.

Since the Hamiltonian $\widehat{H}_t$ is self-adjoint (Hermitian), the operator $\widehat{S}(t)$ is unitary, as a result of which, the state vector norm remains constant (see (3)) and $\widehat{S}\widehat{S}^+ = \hat{I}$, where the superscript '+' denotes the Hermitian adjoint. If the Hamiltonian does not depend on time explicitly $(\widehat{H}_t = \widehat{H}^0)$, then

$$\widehat{S}(t) = \exp\left(-\frac{i}{\hbar}\widehat{H}^0 t\right). \tag{5}$$

Differentiating both parts of the first equality in (2) with respect to time and taking (4) and (5) into account, we obtain the evolution equation of the state vector:

$$i\hbar\frac{d\psi(t)}{dt} = \widehat{H}_t\psi(t), \tag{6}$$

known as the *Schrödinger equation.* Differentiating (3) with respect to t, taking into account (4), and setting $\widehat{H}_t(t) = \widehat{S}^{-1}(t)\widehat{H}_t(0)\widehat{S}(t)$, we obtain the evolution equation for the operator $\widehat{A}_t(t)$ which represents a physical quantity in the Heisenberg representation:

$$\frac{d\widehat{A}_t(t)}{dt} = \frac{\partial\widehat{A}_t(t)}{\partial t} + \frac{1}{i\hbar}[\widehat{A}_t(t), \widehat{H}_t(t)], \tag{7}$$

where

$$[\widehat{A}_t(t), \widehat{H}_t(t)] = \widehat{A}_t(t)\widehat{H}_t(t) - \widehat{H}_t(t)\widehat{A}_t(t)$$

is the commutator of the operators $\widehat{A}_t(t)$ and $\widehat{H}_t(t)$ while $\partial/\partial t$ denotes differentiation with respect to the time parameter t on which the operator $\widehat{A}_t(t)$ depends explicitly. If there is no explicit dependence of the dynamical variable A on time $(\widehat{A}_t(t) = \widehat{A}(t))$, the dynamical equation becomes

$$i\hbar\frac{d\widehat{A}(t)}{dt} = [\widehat{A}(t), \widehat{H}_t(t)], \tag{8}$$

In order to make the dynamical equations in the Heisenberg representation more concrete, it is necessary to give the commutation rules for the operators. The following commutation rules are postulated as the basic relations for canonically conjugate variables, such as the coordinates $\hat{q}_k$ and the momenta $\hat{p}_k$:

$$[\hat{q}_k, \hat{q}_l] = 0, \quad [\hat{p}_k, \hat{p}_l] = 0, \quad [\hat{q}_k, \hat{p}_l] = i\hbar\delta_{kl}\hat{I}, \tag{9}$$

where δ_{kl} is the Kronecker symbol. These rules are analogous to the classical relations involving Poisson brackets.

Using the canonical commutation rules it is possible to show (Blokhintsev, 1981a) that if $\hat{f}(\hat{p}_k, \hat{q}_l)$ is an integral (analytic) function of the operators $\hat{p}_k$ and $\hat{q}_l$, then

$$[\hat{p}_k, \hat{f}] = -i\hbar \frac{\partial \hat{f}}{\partial \hat{q}_k}, \quad [\hat{q}_k, \hat{f}] = i\hbar \frac{\partial \hat{f}}{\partial \hat{p}_k}. \tag{10}$$

As above the square brackets are used as the symbol of the commutator. Applying this rule to the system's Hamiltonian, which has the form of an integral function $\hat{H} = \hat{H}(\hat{p}_k, \hat{q}_k)$, we can make the Heisenberg equations (8), written for these variables, acquire the form of the Hamilton canonical equations

$$\dot{\hat{p}}_k = -\frac{\partial \hat{H}}{\partial \hat{q}_k}, \quad \dot{\hat{q}}_k = \frac{\partial \hat{H}}{\partial \hat{p}_k}. \tag{11}$$

Therefore, the quantization in the Heisenberg's mechanics consists in the substitution of the dynamic variables in the equations of motion by operators, for which operators the commutation rules are postulated.

In view of (2), substitution of (3) into (1) yields the relations

$$\bar{A}(t) = (\psi(0), \, \hat{A}_t(t)\psi(0)) = (\psi(t), \, \hat{A}_t(0)\psi(t)),$$

whence it is clear that, irrespective of whether the vector of state or its operator is time-dependent, the average value of the dynamic variable with a direct physical sense remains the same. This is an evidence of the equivalence of the two representations discussed above.

In view of (7), one can deduce *Ehrenfest's equations* that define the evolution of the averages of physical quantities:

$$\frac{d\bar{A}_t(t)}{dt} = \overline{\frac{\partial A_t(t)}{\partial t}} + \frac{1}{i\hbar}\overline{[\hat{A}_t(t), \hat{H}_t(t)]}. \tag{11a}$$

If a physical quantity A is not explicitly time-dependent, that is, $\hat{A}_t(t) = \hat{A}(t)$, then the equation for the average values takes the form

$$i\hbar \frac{d\bar{A}(t)}{dt} = \overline{[\hat{A}(t), \hat{H}_t(t)]}. \tag{11b}$$

In particular, if $\hat{A} = (\hat{q}_k, \hat{p}_k)$, where $\hat{q}_k$ and $\hat{p}_k$ are the canonically conjugate coordinates and momenta, we find from (11b) that

$$\frac{d\bar{q}_k}{dt} = \overline{\frac{\partial H}{\partial p_k}}, \quad \frac{d\bar{p}_k}{dt} = -\overline{\frac{\partial H}{\partial q_k}}. \tag{11c}$$

These equations are similar to but not identical with the classical equation because in general, the average value of the Poisson brackets does not equal the Poisson brackets of the average values (for example, the average values of the respective derivatives of the Hamilton functions do not generally equal the derivatives of the average values of the dynamic variables). However, there are particluar cases where H is the quadratically linear form of q_k and p_k and therefore the quantum equations for the averages and the classical equations are identical:

$$\frac{d\bar{q}_k}{dt} = \frac{\partial H(\bar{q},\bar{p})}{\partial \bar{p}_k}, \quad \frac{d\bar{p}_k}{dt} = -\frac{\partial H(\bar{q},\bar{p})}{\partial \bar{q}_k}. \tag{11d}$$

Indeed, suppose F is the quadratically linear function of the physical variables A_i, $i = 1, \ldots, n$:

$$F(A,t) = \sum_{k,j=1}^{n} \alpha_{kj}(t)A_k A_j + \sum_{k=1}^{n} \beta_k(t)A_k + \gamma(t). \tag{11e}$$

Recall that the average $\bar{A}$ of any physical quantity A in the quantum state ψ is by definition equal to

$$\bar{A} = (\psi, \widehat{A}\psi). \tag{11f}$$

Therefore, we obtain from (11e) for fixed k $(k = 1, \ldots, n)$:

$$\frac{\partial F(A,t)}{\partial A_k} = \left\langle \sum_{j=1}^{n} [\alpha_{jk}(t) + \alpha_{kj}(t)]\widehat{A}_j + \beta_k(t) \right\rangle =$$

$$= \left(\psi \left\{ \sum_{j=1}^{n} [\alpha_{jk}(t) + \alpha_{kj}(t)]\widehat{A}_j + \beta_k(t) \right\} \psi \right) =$$

$$= \sum_{j=1}^{n} [\alpha_{jk}(t) + \alpha_{kj}(t)](\psi, \widehat{A}_j \psi) + \beta_j(\psi, \psi) =$$

$$= \sum_{j=1}^{n} [\alpha_{jk}(t) + \alpha_{kj}(t)]\bar{A}_j + \beta_k(t) = \frac{\partial F(\bar{A},t)}{\partial \bar{A}_k}.$$

Here the brackets $\langle \rangle$ mean the quantum-mechanical average (expected value).

If the Hamiltonian can be presented as the sum

$$\widehat{H}_t = \widehat{H}^0 + \widehat{H}^1, \tag{12}$$

where $\widehat{H}^0$ is the energy operator disregarding the interaction between the system's parts and $\widehat{H}^1$ is the operator of this interaction, a third representation of the change of state is applied, which is called the *interaction representation*, or the *Dirac representation*. The state vector and the operators both vary with time in this representation according to the equations

$$i\hbar\frac{d\psi(t)}{dt} = \widehat{H}^1\psi(t), \tag{13}$$

$$i\hbar\frac{d\widehat{A}(t)}{dt} = [\widehat{A}(t), \widehat{H}^0]. \tag{14}$$

The average values of the dynamical variables are here the same as above, as can be verified by substituting equations (13) and (14), expressed in terms of the appropriate evolution operators, into the quadratic form (1). The resulting unitary transformation of the state vectors in a frame of reference where the operators do not vary equals the product of the unitary transformations generated by the operators $\widehat{H}^0$ and $\widehat{H}^1_t$.

As was noted above, two physical variables A and B cannot simultaneously possess certain values in either state if their operators $\widehat{A}$ and $\widehat{B}$ do not commute. When the commutation rules are established, it is possible to derive an inequality which must be satisfied by the mean square deviations ΔA and ΔB of these quantities from their average values $\bar{A}$ and $\bar{B}$ respectively. If the commutation rule is

$$[A, \widehat{B}] = i\widehat{M} \neq 0, \tag{14a}$$

where $\widehat{M}$ is a self-adjoint operator, then

$$(\Delta A)^2(\Delta B)^2 \geq \frac{1}{4}(M)^2; \tag{14b}$$

this inequality is called the *generalized Heisenberg uncertainty relation*. For instance, if we consider the conjugate coordinates and momenta, we have the usual Heisenberg inequality

$$\Delta q_k\Delta p_k \geq \frac{\hbar}{2}. \tag{14c}$$

The above remarks have been of a general nature and have not depended on the choice of the basis in the Hilbert space of states. In order to go over to the numerical expression of vectors and operators, it is necessary to produce a concrete basis, or 'select the physical representation of states'. It stands to reason that the average values defined by the quadratic forms (1) are invariant with respect to any representation, but the dynamical equations will be different, and this makes the integration easier. It is very convenient to employ as the basis the orthonormal eigenvectors of the operators of some characteristic physical variables, for example, coordinates, momenta, energy, etc., which will give the name to the corresponding representations.

Besides the common notation of functions, authors of publications on quantum mechanics use the Dirac brackets: $|a\rangle$, the ket-vector; $\langle a|$, its dual bra-vector; $\langle b|a\rangle$, the scalar product; $A|a\rangle$, the product of the operator A and the ket-vector $|a\rangle$; $\langle b|A|a\rangle$, the bilinear form; $\langle x|a\rangle = \psi_a(x)$ is the wave function presenting the state a in the coordinate x-representation (a is called the state index and x is the representation index).

If the wave function $\langle m|a\rangle$ defines a state a in an m-representation, the transition to an r-representation is carried out by means of the transformation

$$\langle r|a\rangle = \sum_m \langle r|m\rangle \, \langle m|a\rangle, \tag{15}$$

which is unitary. The summation index can be either discrete or continuous.

The diadic products $|F_m\rangle\langle F_n|$ of the orthonormal eigenvectors of the operator of a physical quantity F constitute a complete system of operators, and any operator $\widehat{A}$ can be decomposed into them:

$$\widehat{A} = \sum_{m,n} A_{mn}|F_m\rangle \, \langle F_n|. \tag{16}$$

The coefficients of the decomposition

$$A_{mn} = \langle F_m|A|F_n\rangle \tag{17}$$

are called the matrix elements of the operator $\widehat{A}$ in the F-representation. As was shown by Fock in the Appendix to the Russian translation of Dirac (1958), the transition from one representation to another is analogous to the contact transformation in classical Hamiltonian mechanics.

In the representation corresponding to a dynamic variable, the operator of the variable acquires the form of a diagonal matrix. This can be a guide in the choice of a concrete representation.

It is appropriate to note here that in quantum mechanics, depending on the problem and the terms in which it is formulated, it is often convenient to take a particular representation of the state functions (the Schrödinger, Heisenberg, interaction, occupation number representation, etc.) and the representation of independent variables (coordinates, momenta, energy, etc.). Similarly, when solving problems of control of quantum processes or systems, an essential role can be played by a convenient representation of the variables of the problem.

Now let us consider, for instance, the implementation of the commutation rule (9) in the coordinate q-representation. It is necessary and sufficient for this rule to be valid that the momentum operator should have the form

$$\hat{p} = -i\hbar \frac{\partial}{\partial q}. \tag{18}$$

The sufficiency can be verified by the application of the operator

$$\hat{q}\hat{p} - \hat{p}\hat{q} = -i\hbar \left[q \frac{\partial}{\partial q} - \frac{\partial}{\partial q} q \right]$$

to an arbitrary state function ψ with due regard to the sequence of operations. The necessity can be proved (Mackey, 1962) on the basis of the theorems by Stone and von Neumann on the decomposition of a Hilbert space into a direct sum of subspaces.

Any given physical variable A in quantum mechanics can generally take not any values but the values λ belonging to the spectrum of the variable's operator $\widehat{A}$, and these values are eigenvalues of the operator $\widehat{A}$ of the physical variables, that is, $\widehat{A}\psi_k = \lambda_k \psi_k$, where ψ_k is an eigenfunction (element) corresponding to the eigenvalue λ_k. The complete set of the values λ which the variable A can take is called the *spectrum* of the given physical variable. In functional analysis, the spectrum is the set of values of the parameter λ for which the resolvent $R_\lambda = (A - \lambda E)^{-1}$ is not defined. For absolutely continuous self-adjoint operators, these definitions are both practically equivalent because the complete spectrum is the closure of the discrete set of the spectrum of the eigenvalues. Such a spectrum Λ can be discrete or continuous. The self-adjoint (Hermitian) operators associated with the observed physical variables possess only real eigenvalues. An eigenvalue λ_k can be associated with several eigenfunctions $\psi_k^1, \ldots, \psi_k^m$. If this is the case, it is said that λ_k is 'm-tuple', or 'm-fold', and this value is called *degenerate*.

Eigenfunctions ψ_k and ψ_n that correspond to different eigenvalues λ_k and λ_n ($\lambda_k \neq \lambda_n$) of the given operator $\widehat{A}$ are orthogonal. Eigenfunctions $\psi_k^1, \ldots, \psi_k^m$ that correspond to a degenerate eigenvalue λ_k can be chosen so as to be orthogonal to one another. Every eigenfunction of a discrete spectrum, as an element of a Hilbert space, is usually normalized to unity, and every eigenfunction (square-summable function) of the continuous spectrum is normalized to the Dirac delta function.

Thus the whole set $\{\psi_k\}$ of eigenfunctions of the operator $\hat{A}$ of a given physical variable A is a complete orthonormal system of functions into which any state function of the given physical system can be decomposed. To determine the probability of the observed physical variable taking a given value $\lambda_k \in \Lambda$ when the system is in a given state ψ, one can decompose the given state ψ into the complete orthonormal system of eigenfunctions of the operator $\hat{A}$. Indeed, when a certain representation is chosen, that is, when the state vector is decomposed into the orthonormal eigenfunctions of the operator of the dynamic variable A, the components of the decomposition $\langle r|a \rangle$ are probabilistic: the square of the absolute value of the scalar product $\langle r|a \rangle$ is the conditional probability $P(r|a)$ (or the probability density) of observing the value r in the state a:

$$|\langle r|a \rangle|^2 = P(r|a). \tag{19}$$

In particular, $|\psi_a(q)|^2 = P(q|a)$ is the probability density that the quantum-mechanical system in the state a can be observed at a point q. This is why the values of the form $\langle r|a \rangle = \psi_a(r)$ are also called the *probability amplitudes* of observing r at the state a, or simply the *amplitudes* (Feynman et al., 1963).

The transformation (15) can be regarded as a tool allowing us to calculate the amplitude of an event that can occur in many different ways. For a mult-stage process, this formula is similar to the classical formula of the Markov chain, which in the limit, as the steps become infinitesimal, becomes the continuous Feynman integral

$$S(x, t, x_0, t_0) = \int dW[x(\tau)] \exp \frac{i}{\hbar} \int_{t_0}^{t} L(\dot{x}(\tau), x(\tau)) d\tau \tag{20}$$

with respect to the Wiener measure $dW[x(\tau)]$. Here $S(x, t, x_0, t_0)$ is the kernel of the integral evolution operator and $L(\dot{x}, x)$ is the classical Lagrangian function. The functional integral is taken over all the trajectories $x(\tau)$ from the point (x_0, t_0) to the point (x, t) (Feynman and Hibbs 1965).

If the Hamiltonian is not explicitly time-dependent ($H_t = H$), which is the case in the absence of varying macroscopic fields, a state vector satisfying the Schrödinger equation (6) can be represented as

$$\psi(t) = \exp\left(-i\frac{E}{\hbar}t\right)\psi. \tag{21}$$

The time-constant normed vector ψ here must satisfy the equation

$$\widehat{H}\psi = E\psi, \tag{22}$$

which defines the eigenvalues E_m and the eigenvectors ψ_m of the energy operator $\widehat{H}$. The states with definite energies $E = E_m$ defined by the vectors

$$|m\rangle = \exp\left(-i\frac{E}{\hbar}t\right)\psi_m,$$

are called *stationary* since in these states, in view of (19), the probability $P(L|m)$ of observing the certain value L of any mechanical variable remains constant. In particular, if the value L in the stationary state m is completely determined at some instant, it will be definite over the whole time interval of the stationary state as well.

When investigating an oscillation system consisting of oscillators (or identical particles) interacting both with one another and with the external fields, it is preferable to use the so-called 'occupation number representation' (or secondary quantization). If this is the case, the stationary states of the individual oscillators (disregarding the interaction) are discrete sequences with definite energy levels. Each basis state of the whole set of oscillators is associated with a sequence of integers $N_1, N_2, \ldots$ indicating the number of elementary oscillators at the first energy level, the number of them at the second level, etc. These are the so-called occupation numbers for energy levels. The state vector C_N consists of the components $C_{N_1}, C_{N_2}, \ldots$ with the multicomponent (vectorial) integer subscript of the state $N \sim N_1, N_2, \ldots$, where the numbers $N_1, N_2, \ldots$ take every possible integer value.

The Schrödinger equation of a system of pairwise interacting oscillators becomes in the occupation number representation

$$i\hbar\frac{d}{dt}C_N(t) = \widehat{H}_t C_N(t) \tag{23}$$

with Hamiltonian

$$\widehat{H}_t = \sum_{m,n} \hat{a}_m^+ H_{mn}^+ \hat{a}_n + \tfrac{1}{2} \sum_{mm',nn'} \hat{a}_m^+ \hat{a}_{m'}^+ W_{mm',nn'}^+ \hat{a}_n \hat{a}_{n'}, \tag{24}$$

where H_{mn}^+ and $W_{mm',nn'}^+$ are the matrix elements of the energy operators of the individual particles and their pairwise interaction and $\hat{a}_n$ and $\hat{a}_n^+$ are the operators of annihilation and creation of elementary excitations. They are so called because the operator $\hat{a}_n$ diminishes by unity the index of the state corresponding to the energy level E_n while the operator $\hat{a}_n^+$ increases this number by unity. These two operators for the elementary excitations of oscillators and, more generally, for particles (bosons) whose wave function is symmetric with respect to pairwise transpositions of them, satisfy the following commutation rule:

$$\hat{a}_m \hat{a}_n^+ - \hat{a}_n^+ \hat{a}_m = \delta_{mn} \hat{I}. \tag{25}$$

A similar rule for particles (fermions) whose wave function is antisymmetric with respect to pairwise transpositions is

$$\hat{a}_m \hat{a}_n^+ + \hat{a}_n^+ \hat{a}_m = \delta_{mn} \hat{I}. \tag{26}$$

The rule (25) can be obtained from (9) by using the fact that for oscillators with mass m and eigenfrequency ω the annihilation and creation operators are defined by the following expressions (Kuriksha, 1973):

$$\hat{a}_k = \frac{m\omega \hat{q}_k + i\hat{p}_h}{\sqrt{2\hbar m\omega}}, \quad \hat{a}_k^+ = \frac{m\omega \hat{q}_k - i\hat{p}_k}{\sqrt{2\hbar m\omega}}. \tag{27}$$

Using these formulas and (10), it is easy to see that if

$$\widehat{H}_t = \widehat{H}_t(\hat{a}_k, \hat{a}_k^+) \tag{28}$$

is an entire function of the operators $\hat{a}_k$ and $\hat{a}_k^+$, then

$$[\hat{a}_k, \widehat{H}_t] = \frac{\partial \widehat{H}_t}{\partial \hat{a}_k^+}. \tag{29}$$

It follows that the Heisenberg equations (8) for the operators of annihilation and creation of bosons are

$$i\hbar \frac{d\hat{a}_k}{dt} = \frac{\partial \widehat{H}_t}{\partial \hat{a}_k^+}. \tag{30}$$

A methodical discussion of relaxation processes, that is, quantum processes close to a state of equilibrium (Davydov, 1973), has to take into account the interaction of the system with a thermostat, for example, the crystal lattice of a solid. However, if we are not required to analyze the details of the relaxation mechanism, we can limit ourselves in many cases of practical interest to a phenomenological model, in which the relaxation is accounted for by the introduction of an additional dissipation term $\eta_k \hat{a}_k$ into the dynamical equation:

$$\frac{d\hat{a}_k}{dt} + \eta_k \hat{a}_k = -\frac{i}{\hbar} \frac{\partial \widehat{H}_t}{\partial \hat{a}_k^+}, \tag{31}$$

as was done, for instance, in Araujo and Rezende (1974).

The classical Hamiltonian of a particle in an external potential field $V_t(q)$ has the form

$$H_t = \frac{p^2}{2m} + V_t(q).$$

In the coordinate q-representation, we obtain, in view of (18), the energy operator

$$\widehat{H}_t = -\frac{\hbar^2}{2m} \frac{\partial^2}{\partial q^2} + V_t(q), \tag{32}$$

whose substitution into the general Schrödinger equation (6) leads to the well-known wave equation (Landau and Lifshits, 1963)

$$i\hbar \frac{\partial \psi}{\partial t} = -\frac{\hbar^2}{2m} \frac{\partial^2 \psi}{\partial q^2} + V_t(q)\psi. \tag{33}$$

If the potential energy does not depend on time, the corresponding equation for stationary states (22) has the form of the wave equation

$$\frac{\hbar^2}{2m} \frac{\partial^2 \psi}{\partial q^2} + [E - V(q)]\psi = 0. \tag{34}$$

If $V(q)$ does not vary much at distances of the order of the de Broglie wavelength $\lambda = 2\pi\hbar/\sqrt{2mE}$, then a reasonably effective technique of approximate solution of equation (34) is the Wentzel-Kramers-Brillouin (WKB) method, which consists in the construction of consecutive approximations (Flügge, 1971; Blokhintsev, 1981b) for the eikonal $S(q)$ related to the required solution by the formula $\psi = \exp(2\pi i S(q)/\hbar)$. In the zeroth approximation we obtain

$$S_0(q) = \int \sqrt{2m[E - V(q)]}dq.$$

The classical particle changes the sign of its velocity and begins to move in the opposite direction at the point where $V(q) = E$ (the turning point). The quantum-mechanical particle, however, has a finite probability of penetrating through the barrier (the tunnel effect). When the particle passes through the potential barrier in the classically inaccessible domain $q_1 < q < q_2$, where $V(q) > E$, the ψ-function acquires the following multiplier:

$$\exp\left[-\frac{\sqrt{2m}}{\hbar}\int_{q_1}^{q_2}\sqrt{V(q) - E}\,dq\right].$$

Taking into account the first approximation correction

$$S_1(q) = \frac{1}{2}\ln\sqrt{2m[E - V(q)]} - \ln l$$

we obtain particular solutions of equation (34) in the form

$$\psi(q) = \frac{C}{\sqrt[4]{2m[E - V(q)]}}\exp\left\{\frac{i}{\hbar}\int\sqrt{2m[E - V(q)]}\,dq\right\}. \tag{35}$$

The constants of integration can be found from the boundary conditions by means of analytical continuation of this solution over the turning points or else using the asymptotics for the Airy functions.

The concept of the Green's function for various representations of states and independent variables plays an essential role in the descriptions of quantum systems. The Green's function of the Schrödinger equation in the coordinate representation for a system of particles is the unique solution $G(q, t, \xi, \tau)$ of the equation

$$i\hbar\frac{\partial G(q, t, \xi, \tau)}{\partial t} = \widehat{H}\left(q, \frac{\partial}{\partial q}, t\right)G(q, t, \xi, \tau) + i\hbar\delta(q - \xi)\delta(t - \tau)$$

with the initial condition $G(q, t_0, \xi, \tau) = 0$ and any ξ and τ. The Green's function has the sense of the amplitude of the system's transition from the point ξ where it was at the instant τ to the point q at the instant t. If the system was in the state $\psi_0(q)$ at the initial instant $t = t_0$, its state at any t is given by

$$\psi(q, t) = \int G(q, t, \xi, t_0)\psi_0(\xi)\,d\xi.$$

The Green's function makes it possible to give descriptions by means of integral equations, which in many cases is useful for theoretical analysis and for the development of approximate (for example, iterative) methods and numerical calculations.

In fact, let

$$\widehat{H} = \widehat{H}_0\left(q, \frac{\partial}{\partial q}, t\right) + \widehat{H}_1\left(q, \frac{\partial}{\partial q}, t\right)$$

and suppose that the Green's function $G(q, t, \xi, \tau)$ corresponding to the non-perturbed part of H_0 is known. Then it is easy to show that the state of the perturbed system is described by the integral equation

$$\psi(q,t) = \int_{t_0}^{t}\!\!\int G_0(q,t,\xi,\tau)\widehat{H}_1(\xi, \frac{\partial}{\partial \xi}, \tau)\psi(\xi,\tau)d\xi\, d\tau +$$

$$+ \int G_0(q,t,\xi,t_0)\psi_0(\xi)d\xi. \qquad (35a)$$

In particular, setting $\psi(\xi, \tau) = \psi_0(\xi, \tau)$ in the right hand side of this equation, where $\psi_0(\xi, \tau)$ is the non-perturbed solution, we can obtain the first (Born) approximation $\psi'(q, t)$ for the solution $\psi(q, t)$ in explicit form.

Note that, as shown in Butkovskiy (1977), the Green's functions serve to represent complicated controlled distributed systems (in particular, quantum ones) as structural schemes consisting of separate elementary blocks interacting with one another and with the external (environmental) medium, each of the blocks being described by its own Green's function. In a sense, as discussed in Butkovskiy (1977), these structural schemes are quantitative interpretations of the Feynman diagrams.

Until now we have been examining systems with complete (maximum possible) information, that is, those for which one can reliably indicate the quantum state of the system at the initial instant. In these systems with a great number of identical objects, one needs to set definite initial states for the whole ensemble. If the wave function of the ensemble in question is well-determined at the initial instant, it is possible to use it for the description of such an ensemble later on and determine the average values according to (1). A state that can be described by a wave function is called a *pure state*, and the ensemble containing a large number of non-interacting copies in the same pure state, whose averages can be calculated by (1), is called a *pure ensemble*. Pure states are sometimes called simply *quantum-mechanical states*.

If the data on the initial quantum-mechanical state is incomplete, when only the probability that the state occurs is known, then one deals with so-called *mixed states*. Statistical ensembles in which identical copies of the system can occur probabilistically in various quantum relations, are called *mixed ensem-*

bles. The mixed state that characterizes a system with incomplete quantum-mechanical information is defined by means of the statistical operator $\hat{\rho}$, which is usually given by the so-called 'density matrix' which has the form

$$\rho(q,q') = \sum_k w_k \psi_k(q)\psi_k^*(q'), \tag{36}$$

in the coordinate representation, where $w_k \geq 0$ is the probability of observing the system in a state with the wave function $\psi_k(q)$. Since $\sum_k w_k = 1$ and $\int |\psi_k(q)|^2 dq = 1$, we obtain

$$\mathrm{Sp}\,\hat{\rho} = \int \rho(q,q)dq = 1.$$

Having determined the operator $\hat{\rho}$, we can see that it is Hermitian and positive. When $\rho = \psi_l(q')\psi_l(q)$ we have a pure state l as a particular case.

The average value of a dynamical variable A in a mixed state can be found as the result of consistent use of purely quantum-mechanical and statistical averaging. The final expression for the averaging in the mixed state is

$$\langle A \rangle = \iint A(q,q')\rho(q',q)dq'dq = \mathrm{Sp}(\widehat{A}\hat{\rho}). \tag{37}$$

It is essential that the trace of the product of the operators does not vary when they are cyclically permuted, and is invariant with respect to unitary transformations of the state space, that is, the choice of the representation.

Using the Schrödinger equation (6), it can be shown that the evolution of the mixed state can be described by the quantum Liouville equation

$$i\hbar\frac{d\hat{\rho}}{dt} = [\widehat{H}_t, \hat{\rho}]. \tag{38}$$

The evolution of a statistical operator with time from its initial state $\hat{\rho}(0)$ by means of the evolution operator $\widehat{S}(t)$ satisfying equation (4) and the initial condition (5) can be represented in the form

$$\hat{\rho}(t) = \widehat{S}(t)\hat{\rho}(0)\widehat{S}^{-1}(t). \tag{39}$$

As to the initial state, it is often considered to be, for instance, the statistical equilibrium defined by the canonical Gibbs distribution $u_k = \exp[(F - E_k)/\theta]$, where θ is temperature expressed in energy units and F is the free energy.

This distribution is associated with the statistical operator

$$\hat{\rho}(0) = \exp \frac{F - \widehat{H}_0}{\theta}. \tag{40}$$

In consequence, the non-linear response of the quantum system can be represented in the somewhat compact form

$$\hat{\rho}(t) = \widehat{S}(t) \exp \frac{F - \widehat{H}_0}{\theta} \widehat{S}^{-1}(t). \tag{41}$$

Here

$$\widehat{S}(t) = \exp\left[-i\frac{\widehat{H}_0 t}{\hbar}\right] \widehat{P} \exp\left[\frac{1}{i\hbar} \int_{-\infty}^{t} \widehat{H}_\tau^1(\tau)d\tau\right]. \tag{42}$$

The symbol $\widehat{P}\exp(\ldots)$ is the so-called $\widehat{P}$-*exponent*, where $\widehat{P}$ is the Dyson time ordering operator, which sets the product of the time-dependent operators in the chronological order. Here the Hamiltonian is chosen as the sum of the non-perturbed stationary part $\widehat{H}_0$ and the perturbing non-stationary additional part $\widehat{H}_t^1$.

In the linear case of small perturbations in bilinear form

$$\widehat{H}_t^1 = -\sum_j \widehat{B}_j F_j(t) \tag{43}$$

of the fields $F_j(t)$ and their conjugate operators $\widehat{B}_j$, it is possible to construct explicit expressions (the Kubo formulas) for the average values in terms of the fields $F_j(t)$. One of the most widely-used variants of such formulas involving the Green's function with the Heaviside discontinuity factor θ in the form

$$\ll A(t)B(t') \gg = \theta(t - t')\frac{1}{i\hbar}\langle[\widehat{A}(t), \widehat{B}(t')]\rangle, \tag{44}$$

is given by

$$\langle A \rangle = \langle A_0 \rangle - \sum_j \int_{-\infty}^{\infty} \ll A(t)B_j(t') \gg F_j(t')dt'. \tag{45}$$

The average values in the non-linear problem can be expressed as a series in higher order Green's functions, the latter being defined from the system of interrelated equations. We shall not dwell further on this and other approximate techniques of solving problems of quantum statistical mechanics. The material presented above is sufficient for a discussion of the fundamental problem of control of quantum processes and systems. Other concepts will be called upon as required.

We now derive a system of ordinary differential equations for the time amplitudes in the expansion of a pure state function $\psi(q, t)$ into its orthonormal eigenfunctions $\psi_n(q)$, $n = 1, 2, \ldots$, for the operator A of a physical quantity. Suppose that we have an expansion

$$\psi(q, t) = \sum a_n(t)\psi_n(q), \quad t \geq t_0. \tag{46}$$

Multiplying both sides of (46) by $\psi_n^*(q)$, $n = 1, 2, \ldots$, and integrating them, we arrive at the equalities

$$a_n(t) = \int \psi_n^*(q)\psi(q, t)dq, \quad t \geq t_0.$$

Here $a_n(t)$ is the probability amplitude of observing the system in the state ψ_n at the instant t. By definition, the state function $\psi(q, t)$ satisfies the Schrödinger equation

$$i\hbar\frac{\partial\psi(q, t)}{\partial t} = H(t)\psi(q, t), \tag{47}$$

where $H(t)$ is the appropriate non-stationary Hamiltonian. Substituting (46) into (47), we obtain

$$\sum_n i\hbar\dot{a}_n(t)\psi_n(q) = \sum_n a_n(t)H(t)\psi_n(q). \tag{48}$$

Multiplying both sides of this equality by $\psi_m^*(q)$, integrating them with respect to q, and taking into account the orthonormality

$$\int \psi_m^*(q)\psi_n(q)dq = \delta_{mn}, \ m = 1, 2, \ldots; \ n = 1, 2, \ldots, \tag{49}$$

we obtain

$$i\hbar\dot{a}_m(t) = \sum_n a_n(t) \int \psi_m^*(q)H(t)\psi_n(q)dq. \tag{50}$$

We set

$$H_{mn}(t) = \int \psi_m^*(q)H(t)\psi_n(q)dq. \tag{51}$$

This is called the *matrix element of the Hamiltonian* $H(t)$ in the basis of the orthonormal eigenfunctions of the operator A. Therefore, the system of equations we are seeking for the amplitudes $a_n(t)$, $n = 1, 2, \ldots$, has the form

$$i\hbar\dot{a}_m(t) = \sum_n H_{mn}(t)a_n(t), \ m = 1, 2, \ldots, \tag{52}$$

or, in matrix form,

$$i\hbar\dot{a}(t) = H(t)a(t). \tag{53}$$

In particular, if the operator $H(t)$ has the form

$$H(t) = H_0 + V(t), \tag{54}$$

where H_0 does not depend on time (the system is not perturbed) and the expansion (46) is given in the basis of the eigenfunctions $\psi_n(q)$, $n = 1, 2, \ldots$, of the operator H_0, with simple eigenvalues λ_m, $m = 1, 2, \ldots$, then we have

$$H_{mn}(t) = \int \psi_m^*(q)[H_0 + V(t)]\psi_n(q)dq =$$
$$= \int \psi_m^*(q)H_0\psi_n(q)dq + \int \psi_m^*(q)V(t)\psi_n(q)dq =$$
$$= \lambda_m\delta_{mn} + V_{mn}(t), \tag{55}$$

where

$$V_{mn} = \int \psi_m^*(q)V(t)\psi_n(q)dq, \quad m = 1, 2, \ldots; \ n = 1, 2, \ldots, \tag{56}$$

is the matrix element of the perturbation operator $V(t)$ in the basis $\psi_m(q)$ of the orthonormal eigenfunctions of the non-perturbed energy operator H_0.

Therefore, in this concrete case of the operator (54), the system of equations for the amplitudes takes the form

$$i\hbar\dot{a}_m(t) = \lambda_m a_m(t) + \sum_n V_{mn}(t)a_n(t), \ m = 1, 2, \ldots, \tag{57}$$

or, in vector-matrix form,

$$i\hbar\dot{a}(t) = \Lambda a(t) + V(t)a(t), \tag{58}$$

where the matrix $\Lambda = (\lambda_m\delta_{mn})$ is diagonal and $V(t) = (V_{mn}(t))$.

If the perturbing operator depends on the scalar control function $u(t)$ (the control) in the form

$$V(t) = u(t)W, \tag{59}$$

where the operator W does not depend on the control u or time t, the system proves to be a 'bilinear control system' of the form

$$i\hbar\dot{a}_m(t) = \lambda_m a_m(t) + u(t)\sum_n W_{mn}a_n(t), \quad m = 1, 2, \ldots, \tag{60}$$

or, in matrix form,

$$i\hbar \dot{a}I(t) = \Lambda a(t) + u(t)Wa(t), \tag{61}$$

where $W = (W_{mn})$. If we introduce new variables $c_m(t)$, $m = 1, 2, \ldots$, via the formula

$$a_m(t) = c_m(t)e^{(\lambda_m/i\hbar)t} \tag{62}$$

and substitute (62) into (57), then we obtain

$$i\hbar \dot{c}_m(t)e^{(\lambda_m/i\hbar)t} + i\hbar \frac{\lambda_m}{i\hbar} c_m(t)e^{(\lambda_m/i\hbar)t} =$$
$$= \lambda_m c_m(t)e^{(\lambda_m/i\hbar)t} + \sum_n V_{mn}(t)c_n(t)e^{(\lambda_n/i\hbar)t}. \tag{63}$$

Hence

$$i\hbar \dot{c}_m(t)e^{(\lambda_m/i\hbar)t} = \sum_n V_{mn}(t)c_n(t)e^{(\lambda_n/i\hbar)t}, \tag{64}$$

or

$$i\hbar \dot{c}_m = \sum_n V_{mn}(t)e^{((\lambda_n - \lambda_m)/i\hbar)t}c_n(t) = \sum_n V_{mn}e^{i\omega_{mn}}c_n(t), \tag{65}$$

where $\omega_{mn} = (\lambda_m - \lambda_n)\hbar$ is the Bohr frequency of transitions from the energy level $E_m = \lambda_m$ to the energy level $E_n = \lambda_n$.

In order to see more clearly the meaning of the variables $c_m(t)$, we substitute (62) into (46). This yields

$$\psi(q,t) = \sum_n c_n(t)e^{(\lambda_n/i\hbar)t}\psi_n(q) = \sum_n c_n(t)\phi_n(t,q), \tag{66}$$

where

$$\phi_n(t,q) = e^{(\lambda_n/i\hbar)t}\psi_n(q), \quad n = 1, 2, \ldots \tag{67}$$

It is easy to see that (67) is the solution of the Schrödinger equation corresponding to the nth stationary state

$$i\hbar \frac{\partial \phi}{\partial t} = H_0 \phi. \tag{68}$$

Therefore, as is clear from (66), $c_n(t)$ is the coefficient of the expansion of the state $\psi(q,t)$ in the stationary states of the given system with Hamiltonian H_0.

In conclusion, here is a review of the basic principles underlying non-relativistic quantum mechanics. Physical quantities (observables) in quantum mechanics are associated with self-adjoint operators in a Hilbert space $\mathcal{H}$ whose concrete realization is defined by the method of choosing the representation of the operators. For the operators of momenta $\hat{p}_m$ and coordinates $\hat{q}_n$, the following commutation rule is postulated

$$[\hat{p}_m, \hat{q}_n] = \hat{p}_m \hat{q}_n - \hat{q}_n \hat{p}_m = \frac{\hbar}{i} I \delta_{mn}, \tag{69}$$

where $i = \sqrt{-1}$, $\hbar$ is the Planck's constant, I is the identity operator, and δ_{mn} is the Kronecker symbol. This rule gives the commutation relations for other observables that are functions of the momenta and the coordinates.

The state of a quantum system with maximum (complete) information, called a *pure state*, is defined by the elements (vectors) $|\psi\rangle$ in the Hilbert space H in which the observable operators act. The elements of the conjugate space H^* are denoted by $\langle\psi|$.

A *mixed*, that is, non-deterministic state of the quantum system is given using the so-called 'statistical' operator $\rho = \rho(t)$. This is a positive self-adjoint kernel operator with trace $\mathrm{Sp}\,\hat{\rho} = 1$.

The evolution postulate. When there is no interaction between the quantum system and the macroscopic measuring instrument but there are external classical fields, the statistical operator varies according to the law

$$\hat{\rho}(t + t_0) = \widehat{U}(t)\hat{\rho}(t_0)U^{-1}(t), \tag{70}$$

where $\widehat{U}(t)$ is the unitary operator of evolution satisfying the Schrödinger equation

$$i\hbar\frac{\partial \widehat{U}}{\partial t} = \widehat{H}\widehat{U}, \ \widehat{U}(0) = \hat{I}. \tag{71}$$

Here $\widehat{H} = \widehat{H}(\cdot)$ is the self-adjoint energy operator, the Hamiltonian of the system, which is generally independent of the external fields regarded as the control u.

The process of measuring physical quantities in quantum mechanics is characterized by the following postulates.

The postulate of observables. Any result a of measuring the observable A always belongs to the spectrum of the eigenvalues of the associated operator $\widehat{A}$. If the system is in the state $\hat{\rho}(t)$ at the instant of measurement, then the a priori probability of a being within the spectral interval Δ equals

$$p(a \in \Delta|t) = \mathrm{Sp}[\hat{\rho}(t)\hat{\pi}(\Delta)], \tag{72}$$

where $\hat{\pi}(\Delta)$ is the projection operator onto the subspace $L_A \subset \mathcal{H}$ of the eigen-vectors of the operator $\hat{A}$ corresponding to the indicated spectral interval.

The postulate of reduction. At the instant t, when the result of mea-suring is read, there occurs from the point of view of the observer a reduction of the statistical operator according to the rule

$$\hat{\rho}(t) \rightarrow \hat{\rho}_a(t) = \frac{\hat{\pi}(a)\hat{\rho}(t)\hat{\pi}(a)}{\mathrm{Sp}[\hat{\pi}(a)\hat{\rho}(t)\hat{\pi}(a)]}, \tag{73}$$

where $\hat{\pi}(a)$ is the projector onto the subspace $L \subset \mathcal{H}$ of the eigenvectors of the operator $\hat{A}$ corresponding to the eigenvalue a. For instance, if L_a is a one-dimensional subspace, then the system goes over to the pure state $\hat{\rho}_a = |a\rangle\langle a|$, where $|a\rangle$ is the eigenvector of the operator $\hat{A} : \hat{A}|a\rangle = a|a\rangle$. In the general case, the reduction of the pure state occurs as a result of joint measurement of observables forming a complete set of simultaneously measurable quantities.

Therefore, if the Hamiltonian H is a deterministic operator, the incom-pleteness of the information on the system's state is eliminated after the first measurement; however, owing to evolution of the state, a repeated measure-ment of the quantity A generally does not yield the initial result. In compliance with (72), the probability of observing the result a again after the time t equals $\langle a|U(t)|a\rangle$. One obtains an identical result of measurement with probability 1 if for some τ (and therefore also $\forall n\tau$, $n \in \mathcal{L}$)

$$[\hat{A}(\tau), \hat{A}(0)] = 0, \tag{74}$$

where $\hat{A}(\tau) = \hat{U}^{-1}(\tau)\hat{A}(0)\hat{U}(\tau)$. The case of periodically reliable measurements of an observable and the possibility of continuous non-perturbing measurement of special integrals of motion were discussed by Moncrieff (1978).

Note that similar results can also be obtained for a wider class of observ-ables if the quantum system is controllable. The validity of (74) can be due to the synthesis of the desired evolution operator $\hat{U}(\tau)$. In periodically non-stationary systems such possibilities are provided by special properties of the dynamical symmetry (Malkin and Mahko, 1979).

Controllability and Finite Control of Dynamical Systems

The purpose of this appendix is to present very briefly, without any detailed deductions and proofs, the basic results in the analytic theory of controllability and finite control employed in the text of this book. If needed, the details can be found elsewhere (Andreev, 1976; Butkovskiy, 1975, 1977; Voronov, 1979; N. Krasovskii, 1968; Feldbaum and Butkovskiy, 1971).

In the theory of control of dynamic systems, the notion of 'controllability' is a key one along with other vital notions such as, for instance, 'stability'. The problem of stability, first posed by R. Kalman in 1960, consists in the solution of problems on the possibilities in principle of realizing a control for taking a given controlled system from one state (or set of states) to another state (or another set of states). On the whole, this problem still seems to be far from being completely solved. The only case that can be said to be almost complete is perhaps the development of the controllability problem for finite-dimensional linear stationary lumped-parameter systems whose motion can be described by a finite system of ordinary linear differential equations with constant coefficients.

This problem is still very difficult for non-linear systems described by ordinary differential equations and *a fortiori* for linear and non-linear systems described by partial differential equations.

The problem of finite control was posed somewhat later, first for distributed-parameter systems in connection with the optimal control problem (Feldbaum and Butkovskiy, 1971). The essence of the finite control problem is to find at least one control function or set of control functions *realizing* the conditions of controllability of the given dynamic system. As a rule, the problem of taking the system from one state to another is posed for a finite time interval $\left[t_0, t_1\right]$ and, for stationary systems, for an interval $[0, T]$, $T \geq 0$. Therefore, without loss of generality, the control functions realizing controllability can be regarded as finite functions of time with support $\left[t_0, t_1\right]$ or $\left[0, T\right]$. Hence the natural term *finite control*.

A2.1. Controllability and Finite Control of Linear Finite-Dimensional Systems

These systems are described by the equation

$$\dot{q}(t) = Aq(t) + f(t) + Bu(t), \tag{1}$$

where $q = (q_1, \ldots, q_n)$ is the state, $u = (u_1, \ldots, u_r)$ is the control, $f = (f_1, \ldots, f_n)$ is a given function describing the external perturbations acting upon the system, and A and B are constant matrices of appropriate dimensions. The *problem of controllability* consists in finding conditions to be imposed on the parameters of the system (1), that is on A and B, under which for any two points q_0 and q_* there exists a control $u = u(t)$, $0 \leq t \leq T$, taking the system (1) from the given initial state $q(0) = q_0$ to the given final state $q(T) = q_*$ for a finite $T \geq 0$. The solution of this problem is given by the following theorem due to Kalman.

Theorem. *For any $T \geq 0$, the solution of the controllability problem exists if and only if the matrix $(B, AB, \ldots, A^{n-1}B)$ has rank n.*

Suppose that the system (1) is such that the conditions of the Kalman theorem are satisfied. Then the finite control problem consists in describing the entire set of controls $(0 \leq t \leq T)$ that take the system (1) from q_0 to q_*. This problem can be solved in several different ways. We shall describe here two methods of its solution. The first method is very specific and relates to a certain form of the system (1). The second method is more general and can be used for the solution of problems of finite control in distributed-parameter systems.

First method of solution. This is based on the reduction of the finite control problem to a system of integral equations or, in other words, the reduction to a certain moment problem. In order to carry out this reduction, we write the solution $q(t)$ of equation (1) with the initial condition $q(0) = q_0$:

$$q(t) = e^{At}q_0 + e^{At}\int_0^t e^{-A\tau}f(\tau)d\tau + e^{At}\int_0^t e^{-A\tau}Bu(\tau)d\tau. \qquad (2)$$

We set $t = T$ in (2). Then in view of the final condition $q(\tau) = q_*$ we have

$$q_* = e^{AT}q_0 + e^{AT}\int_0^T e^{-A\tau}f(\tau)d\tau + e^{AT}\int_0^T e^{-A\tau}Bu(\tau)d\tau.$$

Hence we obtain the required moment problem

$$\int_0^T e^{-A\tau}Bu(\tau)d\tau = \alpha, \qquad (3)$$

where

$$\alpha = -q_0 - \int_0^T e^{-A\tau}f(\tau)d\tau + q_* e^{-AT} \qquad (4)$$

is the given vector. Since the system of integral equations (3) is linear with respect to the sought function $u(t)$, $0 \le t \le T$, its general solution is known to be the sum of two solutions: the general solution $u_0(t)$ of the corresponding homogeneous equation

$$\int_0^T e^{-A\tau}Bu(\tau)d\tau = 0 \qquad (5)$$

and a particular solution of the inhomogeneous equation (3).

The control $u_0(t)$, $0 \le t \le T$, is called the *null-finite* because it transfers the system (1), where we set $f(t) = 0$, from the zero initial state $q(0) = 0$ to the zero state $q(T) = 0$ again. Since $u_0(t)$ is required as a finite function with support $[0, T]$, equation (5) can be written with infinite limits:

$$\int_{-\infty}^{\infty} e^{-A\tau}Bu(\tau)d\tau = 0. \qquad (6)$$

We confine ourselves to the case when u is a scalar. Then it is possible to show that the null-finite control has the form

$$u_0(t) = \Delta(p)\gamma(t), \qquad (7)$$

where $p = d/dt$ is the differentiation operator, $\Delta(p) = |pE - A| = a_0 p^n + a_1 p^{n-1} + \ldots + a_n$ is the characteristic polynomial of (1), and $\gamma(t)$ is an arbitrary finite function with support $[0, T]$ and possessing n generalized derivatives.

The fact that (7) satisfies (6) can be easily verified by substituting (7) into (6) and integrating by parts taking into account the fact that because of finiteness, the function $\gamma(t)$ and all its derivatives vanish identically outside $[0, T]$. It is also necessary here to bear in mind that the matrix A is, according to the Cayley-Hamilton theorem, a solution of the equation $\Delta(A) = 0$.

We now seek a particular solution $u_*(t)$ of the system (3) in the form

$$u_*(t) = [\delta(t), \delta'(t), \ldots, \delta^{(n-1)}(t)]H^{-1}\alpha, \tag{8}$$

where $H = (B, AB, \ldots, A^{n-1}B)$ is the Kalman matrix. Since it has been assumed that the system (1) is controllable, there exists the inverse matrix H^{-1} in compliance with the Kalman theorem. Therefore, when u is scalar, the required set of all finite controls can be described by the formula

$$u(t) = u_0(t) + u_*(t) = \Delta(p)\gamma(t) + [\delta(t), \ldots, \delta^{(n-1)}(t)]H^{-1}\alpha. \tag{9}$$

We now give a second more general method of solving the posed problem of finite control.

Second method of sulution. The Paley-Wiener-Schwartz theorem is vital for this method.

Paley-Wiener-Schwartz Theorem. *In order that a finite function $\gamma(t)$ with support $[-\tau, \tau]$ be representable in the form*

$$\gamma(t) = r(t) + \sum_{i=1}^{s-1} \sum_{k=1}^{m} c_{ik} \delta^{(i)}(t - t_k), \tag{10}$$

where $r(t)$ is a finite square integrable function on $[-\tau, \tau]$ and the c_{ik} are constants, it is necessary and sufficient that the Fourier transform $\tilde{\gamma}(\omega)$ of $\gamma(t)$ (1) does not grow (in absolute value) faster than $|\omega|^3$ and (2) can be extended from the ω-axis to the entire complex plane $z = \omega + j\xi$ as an entire (analytic) function of degree τ.

Recall that a function $F(z)$ is called entire if it can be represented on the entire finite part of the z-plane as a convergent power series

$$F(z) = a_0 + a_1 z + \ldots + a_n z^n + \ldots \tag{11}$$

Consequently, the entire function $F(z)$ does not have any singularities (for example, poles) in any bounded part of the z-plane.

An entire function $F(z)$ is called a *function of finite order* if $|F(z)| < e^{\alpha|z|}$ for any $|z| > M$, where M is sufficiently great and α is a positive number that does not depend on z. The lower bound τ of numbers α for which this inequality holds is called the *order of the entire function* $F(z)$. Note that the order τ of the entire function $F(z)$ can be calculated by the formula

$$\tau = \lim_{k \to \infty} \sqrt[k]{k!|a_k|}, \tag{12}$$

where a_k, $k = 0, 1, 2, \ldots$, are the coefficients of the series (11).

Before presenting the second method of solving the finite control problem we note that, as is clear from (4), we can set $f(t) = 0$ and $q_* = 0$ without loss of generality. In consequence, it is sufficient to solve the problem of finite control taking the system (1) from an arbitrary initial point $q(0) = q_0$ to the origin $q(T) = 0$. When this problem is solved, it is sufficient in order to arrive at the general case ($f \neq 0$ and $q_* \neq 0$) to do the following: substitute q_0 in the solution so obtained by the expression

$$-q_0 - \int_0^T e^{-At} f(t) dt + q_* e^{-At}.$$

Thus, we suppose in what follows that $q_* = 0$ and $f(t) = 0$. We take the Fourier transform of (1) according to the formula

$$\tilde{f}(i\omega) = \int_{-\infty}^{\infty} e^{-i\omega t} f(t) dt, \ i = \sqrt{-1}. \tag{13}$$

In view of the initial condition, we obtain

$$i\omega \tilde{q}(i\omega) - q_0 = A\tilde{q}(i\omega) + B\tilde{u}(i\omega),$$

or

$$(i\omega E - A)\tilde{q}(i\omega) = B\tilde{u}(i\omega) + q_0.$$

Using the matrix method, we can solve this linear algebraic equation for $\tilde{q}(i\omega)$:

$$\tilde{q}(i\omega) = \frac{C(i\omega)B\tilde{u}(i\omega) + C(i\omega)q_0}{\Delta(i\omega)}, \tag{14}$$

where $C(i\omega)$ is the adjoint of the matrix $(i\omega E - A)$ and $\Delta(i\omega) = |i\omega E - A|$. We extend the expression for $\tilde{q}(i\omega)$ to the entire complex plane, replacing ω by $z = \omega + i\xi$. In order for $\tilde{q}(i\omega)$ to be an analytic function (the corresponding function $q(t)$ will be finite), it is necessary, according to the Paley-Wiener-Schwartz theorem, that this expression have no poles in any finite part of the z-plane. Evidently, these poles can only appear where the denominator $\Delta(iz) = 0$. Therefore, there are no poles if the numerator of this expression becomes zero at these points. This requirement, as will be shown below, completely determines the form of the function $\tilde{u}(iz)$ and therefore, the form of the finite control $u(t)$ being sought.

Indeed, suppose that $z_1, \ldots, z_n$ are simple roots of $\Delta(iz)$ (the case of multiple roots can be considered in a similar way). Then

$$C(iz_k)B\tilde{u}(iz_k) + C(iz_k)\,q_0 = 0, \quad k = 1, \ldots, n. \tag{15}$$

Supposing for simplicity's sake that u is a scalar, we then obtain

$$\tilde{u}(iz_k) = -\frac{C(iz_k)q_0}{C(iz_k)B} = \beta_k, \quad k = 1, \ldots, n. \tag{16}$$

In consequence, the problem is reduced to finding an analytic function of a given order satisfying the interpolation conditions (16). It can easily be shown that the conditions (16) have a concrete meaning and the numbers β_k obtained are defined and finite if the system is controllable. The converse statement is also valid. The general solution of the interpolation problem (16) can easily be found with the aid of the interpolation Lagrange polynomial

$$\tilde{u}(iz) = \sum_{k=1}^{n} \frac{\beta_k \tilde{\phi}_1(iz)}{\tilde{\phi}_1'(iz)(z - z_k)} + \tilde{\phi}_2(iz), \tag{17}$$

where $\tilde{\phi}_1(iz)$ and $\tilde{\phi}_2(iz)$ are arbitrary entire functions of finite orders with zeros at the points $z_1, \ldots, z_n$. For instance, if we set $\tilde{\phi}_1(iz) \equiv \Delta(iz)$ and $\tilde{\phi}_2(iz) = \gamma(z)\Delta(iz)$, where $\gamma(z)$ is an arbitrary analytic function of finite order, then, applying the inverse Fourier transformation to (10), we arrive at the general solution of the finite control problem in the form (16).

A2.2. Finite Control of Linear Distributed Systems

The second method of solving finite control problems (see A2.1), as was mentioned above, is also efficient for linear distributed systems. We illustrate it by

the solution of the problem of quenching the vibrations of a string described
by the one-dimensional wave equation

$$\ddot{Q} = Q'', \quad 0 < x < \pi, \ t > 0, \tag{1}$$

with the boundary conditions

$$Q(0,t) = u(t), \quad t \geq 0, \tag{2}$$
$$Q(\pi,t) = 0, \quad t \geq 0, \tag{3}$$

and the initial conditions

$$Q(x,0) = Q_0(x), \quad 0 \leq x \leq \pi, \tag{4}$$
$$\dot{Q}(x,0) = Q_1(x), \quad 0 \leq x \leq \pi. \tag{5}$$

It is required to find a control $u(t)$, $0 \leq t \leq T$, that quenches the vibrations of
the string:

$$Q(x,T) = 0, \ 0 \leq x \leq \pi, \tag{6}$$
$$\dot{Q}(x,T) = 0, \ 0 \leq x \leq \pi, \tag{7}$$

Applying time Fourier transformation to both sides of equation (1) and
taking into account (4) and (5), we obtain

$$-\omega^2 \widetilde{Q} - j\omega Q_0 - Q_1 = \widetilde{Q}'', \tag{8}$$

where $\widetilde{Q} = \widetilde{Q}(x,\omega)$ is the time Fourier transform for $Q(x,t)$. The boundary
conditions for (8), also subjected to the time Fourier transformation, yield the
conditions

$$\widetilde{Q}(0,\omega) = \tilde{u}(\omega), \tag{9}$$
$$\widetilde{Q}(\pi,\omega) = 0. \tag{10}$$

Solving (8) as an ordinary second-order linear differential equation in x by
standard methods (ω is a parameter) with the boundary conditions (9), we
obtain a definite expression of $\widetilde{Q}(x,\omega)$ in the form of the ratio of two entire
functions, the denominator being the characteristic function $\sin \pi z$. The zeros
of this function are obviously $z_k = k$, $k = 0, \pm 1, \pm 2, \ldots$. The condition that the

numerator of the expression $\widetilde{Q}(x, z)$ vanishes at these points gives the required interpolation conditions

$$\tilde{u}(k) = -\int_0^\pi \sin kx \left[iQ_0(x) + \frac{1}{k}Q_1(x) \right] dx = \beta_k, \tag{11}$$

$$k = 1, 2, \ldots$$

If the conditions (11) hold for $k = 1, 2, \ldots$, then, as is easy to show from the form of $\widetilde{Q}(x, \omega)$, they will also hold for $k = 0, -1, -2, \ldots$. Therefore, the problem has been reduced to finding an entire function $\tilde{u}(z)$ satisfying the conditions (11).

To solve this problem in the general case, we can make use of the Lagrange formula (1.17), setting $n \to \infty$. However, the problem in this case can be solved more easily: it clearly reduces to finding the inverse Fourier transform of the entire function

$$\tilde{u}(\omega) = -\int_0^\pi \sin \omega x \left[iQ_0(x) + \frac{1}{\omega}Q_1(x) \right] dx \tag{12}$$

Finding the inverse Fourier transform of (12), we obtain ($0 \le t \le 2\pi$):

$$u(t) = \frac{1}{2}\bar{Q}_0(t) + \frac{1}{2}\int_0^t \bar{Q}_1(x)dx - \frac{1}{2\pi}\int_0^\pi (\pi - x)Q_1(x)dx, \tag{13}$$

where

$$\bar{Q}_0(t) = \begin{cases} Q_0(t), & 0 \le t < \pi, \\ -Q_0(2\pi - t), & \pi \le t \le 2\pi, \end{cases}$$

$$\bar{Q}_1(t) = \begin{cases} Q_1(t), & 0 \le t \le \pi, \\ -Q_1(2\pi - t), & \pi \le t \le 2\pi \end{cases} \tag{14}$$

Given the finite control (13) carrying out the desired quenching of the vibrations in the system, we can add an arbitrary null-finite control $u_0(t)$ corresponding to $\tilde{u}_0(\omega) = \gamma(\omega) \sin \pi\omega$ ($\gamma(\omega)$ is an arbitrary entire function) which solves the interpolation problem (10) for β_k, $k = 1, 2, \ldots$. For instance, the control

$$u_0(t) = a[\delta(t) - \delta(2\pi - t)],$$

where a is an arbitrary constant, is a null-finite control $u_0(t)$. The wide spectrum of choice of the null-finite control using $\gamma(z)$ makes it possible to select from the large set of all finite controls the one that minimizes some given functional.

We conclude this section by noting that since the expression for the null-finite control contains the multiplier $\sin \omega \pi$, whose order as an entire function is π, it follows that the minimum time for quenching the vibrations is generally no less than 2π. This conclusion has a simple and clear physical sense: given that the control is carried out at one end of a vibrating string, the time of quenching cannot generally be less than the travel time of the perturbation from one end of the string and back. This time in the given case equals 2π (in dimensionless units).

A2.3. A New Differential Geometric Method of Solving the Problems of Finite Control of Non-Linear Finite-Dimensional Dynamical Systems

We shall start by presenting this method for two-dimensional systems that can move on two-dimensional manifolds (a plane, a sphere, a torus, etc.). Numerous examples show that this method is especially illustrative for two-dimensional systems and provides constructive solutions of problems.

Suppose that a controlled process is described by the system in vectorial form

$$\frac{dq}{dt} \equiv \dot{q} = f(q, u), \quad t > 0, \tag{1}$$

where $q = (q_1, q_2) \in \mathbf{R}^2$ is the vector of controlled coordinates on the plane and u is the control, the value of which must belong at every instant t to a given set U. The nature of U can be quite arbitrary. The control $u = \phi(t)$, $t > 0$, where $u \in \mathcal{U}$, will be called an *admissible control*, and the vector $\dot{q} = f(q, u)$, when $u \in \mathcal{U}$, will be called an *admissible velocity vector*.

The problem of finite control consists in establishing, for two given points $q_0 \in \mathbf{R}^2$ and $q_* \in \mathbf{R}^2$, the existence of an admissible control $u = \phi(t)$ that takes the system (1) from the initial point $q(0) = q_0$ to the final point q_* during the finite time $T \geq 0$ or else $q(t) \rightarrow q_*$ as $T \rightarrow \infty$. If such a control exists, it is required to find at least one control function or, if it is not unique, describe the whole set of the controls. The control $u = \phi(t)$, $0 \leq t \leq T$, solving this problem, as was mentioned above, is called a *finite control with support* $[0, T]$.

An essential role in the solution of the problem is played by the set of points (denoted by $f(q, \mathcal{U})$) in the plane $\mathbf{R}^2$; this set is composed of the ends of the vectors $\dot{q} = f(q, u)$ originating from the point $q \in \mathbf{R}^2$ when u runs over the whole set $\mathcal{U}$ of its admissible values. We suppose that $f(q,\mathcal{U})$ is non-empty for any $q \in \mathbf{R}^2$. If the set $f(q,\mathcal{U})$ proves to be non-convex, then we take its convex hull, that is, the smallest convex set containing $f(q,\mathcal{U})$. We call this convex hull the *set of admissible motion velocities of the system* (1) if the system is at the point q (Figure A2.3.1) and again denote it by $f(q,\mathcal{U})$.

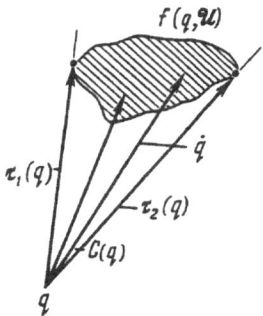

Fig. A2.3.1

This substitution of the generally non-convex set $f(q,\mathcal{U})$ by its convex hull has a definite justification. In fact, since each vector $\dot{q}$ whose end belongs to a convex set can be represented as a linear combination (with positive coefficients) of two admissible direction vectors $\dot{q}_1 = f(q,u_1)$, $u_1 \in \mathcal{U}$, and $\dot{q}_2 = f(q,u_2)$, $u_2 \in \mathcal{U}$, it follows that

$$\dot{q} = \alpha\dot{q}_1 + \beta\dot{q}_2, \quad \alpha \geq 0, \ \beta \geq 0, \ \alpha + \beta = 1. \tag{2}$$

Therefore if the switch of control u from the value u_1 to the value u_2 and back is carried out frequently enough, it is evident that the motion 'on the average' will occur with the required velocity $\dot{q}$ belonging to the convex hull $f(q,\mathcal{U})$.

Consequently, we can assume $f(q,\mathcal{U})$ to be a convex set without loss of generality. It is interesting to note that the system (1), where $u \in \mathcal{U}$, can be described by the following condition, called the *differential inclusion* or the *contingency equation*:

$$\dot{q} \in f(q,\mathcal{U}), \ q(0) = q_0, \ t > 0. \tag{3}$$

In these terms, the finite control problem can be reformulated as follows: find an absolutely continuous function $q = \xi(t)$, $0 \leq t \leq T$, such that $\xi(0) = q_0$ and

$\xi(T) = q_*$, the function being a solution of the differential inclusion (3), that is,

$$\dot{\xi}(t) \in f(\xi(t), \mathcal{U}) \tag{4}$$

for almost all $t \in (0, T)$.

Let us return to the convex set $f(q, \mathcal{U})$. We single out the case when $f(q, \mathcal{U})$ is a domain on $\mathbf{R}^2$, as shown in Figure A2.3.1. Two subcases are possible:

(1) the point q is inside the domain $f(q, \mathcal{U})$ (see Figure A2.3.2);

(2) the point q is outside the domain $f(q, \mathcal{U})$ or on its boundary (Figure A2.3.1).

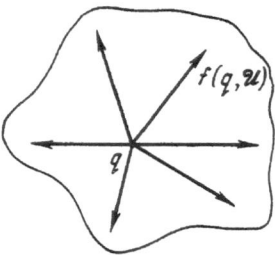

Fig. A2.3.2

In the case (1), motion from q is possible in any direction. If there is a connected domain $\mathcal{D}$ in $\mathbf{R}^2$ at each point q where this case holds, then this domain can be called a *free motion domain*. Suppose that the initial and final points q_0, q_* are both inside $\mathcal{D}$. Then the finite control problem is trivial: for instance, it is possible to get from q_0 to q_* along any trajectory lying entirely inside $\mathcal{D}$. The domain $\mathcal{D}$ of free motion, if it exists, can be identified *a priori*. Obviously, the domain $\mathcal{D}$ is defined by the following condition: the point 0 belongs to $f(q, \mathcal{U})$. The boundary of the domain $\mathcal{D}$ and its equation must also comply with the condition that the point 0 belongs to the boundary of $f(q, \mathcal{U})$.

The second case, where q does not belong to the domain $f(q, \mathcal{U})$ or belongs to its boundary, is more complicated. Below we assume that we are dealing with such a case. Then the set of solutions of the contingency equation (3) forms a so-called *'trajectory funnel'* with vertex at the point q_0. The boundary of this funnel in the two-dimensional case is found very simply: it is the solution of the Cauchy problem for the two differential equations (each of second order):

$$\dot{q} = \tau_1(q), \quad q(0) = q_0, \ t > 0,$$
$$\dot{q} = \tau_2(q), \quad q(0) = q_0, \ t > 0,$$

$$(5)$$

where $\tau_1(q)$ and $\tau_2(q)$ are assumed to be non-zero 'extreme' vectors whose meaning is clear from Figure A2.3.1. These vectors define a cone $C(q)$ with the vertex at the point q. The cone consists of all rays having a non-empty intersection with $f(q, \mathcal{U})$.

The problem of finding the boundary of the trajectory funnel (3), when the dimension of the system (1) is greater than two, is not at all simple. This problem will be discussed at the end of this section where the multidimensional cone $C(q)$ will play an essential role in it.

The idea behind our differential geometric method for solving the problem of finite control consists in the following:

(1) Plot a family of trajectory funnel boundaries in the $\mathbf{R}^2$ plane (or some other two-dimensional manifold), forming a sufficiently dense covering of the plane or some specified part of it.

(2) Hatch the curves of this family of trajectory funnel boundaries at their external sides, that is, opposite to where the set $f(q, \mathcal{U})$ is.

This hatched family of curves on the plane will be referred to as the phase portrait of the system (1) or the inclusion (2). Such a phase portrait evidently allows us to plot a continuous curve s (possessing the tangent almost everywhere) connecting the points q_0 and q_*. As a rule, this curve is not unique. The principal requirement is that these curves must meet the curves of the hatched family in the direction from q_0 to q_* in admissible fashion, that is, from the hatched side only. Let us term such a curve *admissible*. To illustrate the point, let us consider several examples.

1. Take the system $\dot{q}_1 = q_2$, $\dot{q}_2 = u$, $|u| \leq l$, $l > 0$. The equations of the family of trajectory funnel boundaries are $\dot{q}_1 = q_2$, $\dot{q}_2 = \pm l$. Integrating their ratio $dq_1/dq_2 = \pm q_2/l$, we obtain the equation of a family of parabolas

$$q_1 = \pm \frac{1}{2l}q_2^2 + C.$$

This family is plotted and hatched in Figure A2.3.3. The phase portrait of the system in question shows that the system is completely controllable. An admissible trajectory is plotted as well.

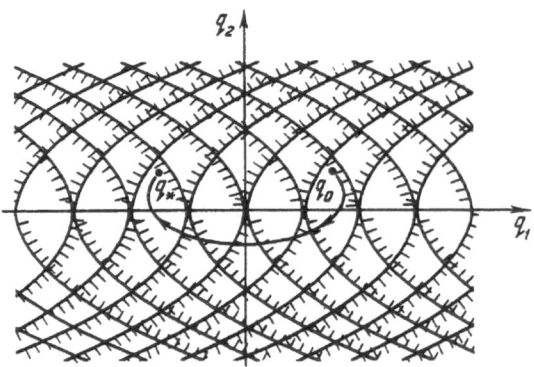

Fig. A2.3.3

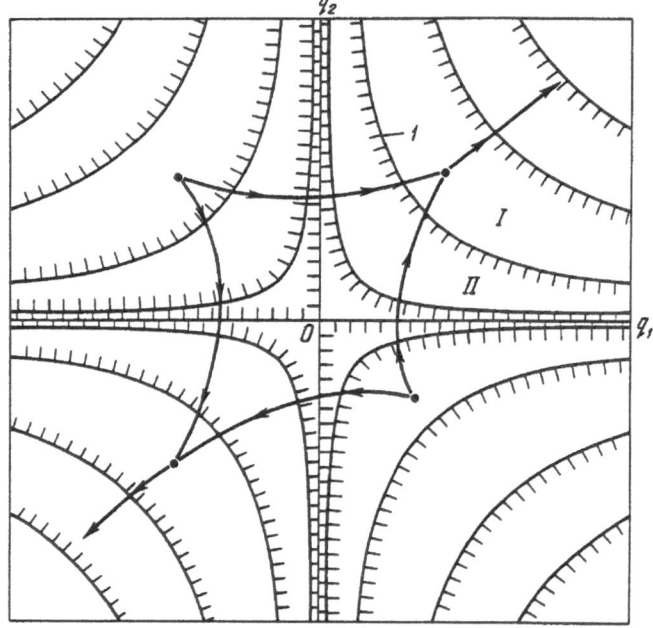

Fig. A2.3.4

2. Suppose that the system is bilinear:

$$\dot{q}_1 = q_2 + uq_1, \quad \dot{q}_2 = q_1 - uq_2, \quad |u| \le l, \; l > 0. \tag{6}$$

The equation of the family of trajectory funnel boundaries is

$$\frac{dq_1}{dq_2} = \frac{q_2 \pm lq_1}{q_1 \mp lq_2}. \tag{7}$$

Assume that u is unbounded, that is, $l \to \infty$. Then (7) becomes

$$\frac{dq_1}{dq_2} = -\frac{q_1}{q_2}.$$

Hence $q_1 q_2 = C$ is a family of hyperbolas that is plotted and hatched in Figure A2.3.4. This figure shows that the system is not completely controllable. From the point q_0 of the first quadrant it is only possible to get to any point 'higher' than the initial hyperbola passing through q_0. The situation is similar in the third quadrant. From the second quadrant it is possible to get to any point of the first and the third quadrants but not the fourth quadrant, etc.

3. Suppose that the bilinear system is $\dot{q}_1 = q_2 + uq_1$, $\dot{q}_2 = -q_1 - uq_2$ where u is unbounded. Then the equation of the family of trajectory funnel boundaries will have the same form as in the preceding example. However, the hatching will be different and, unlike the preceding example, this system is completely controllable, as is clear from Figure A2.3.5.

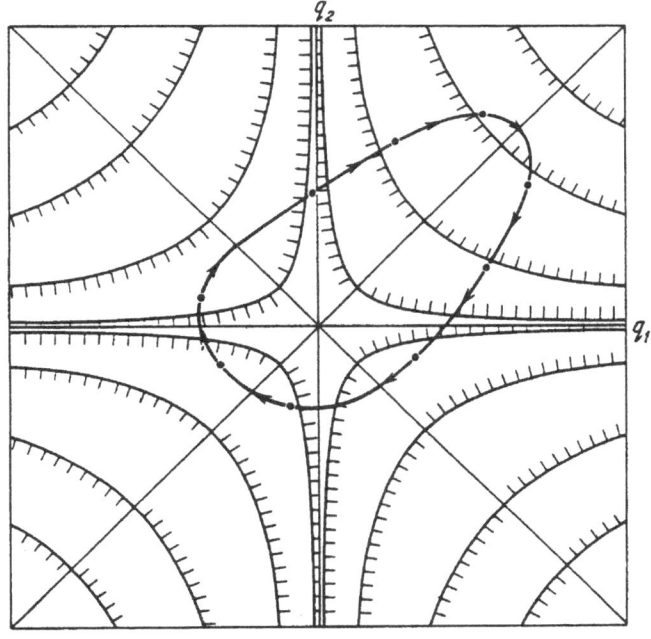

Fig. A2.3.5

4. Suppose that the bilinear system is

$$\dot{q}_1 = -aq_2, \quad \dot{q}_2 = aq_1 + uq_3, \quad \dot{q}_3 = -uq_2, \quad a > 0, \tag{8}$$

where u is unbounded. The equations define rotational motions in both classical and quantum mechanics (the Bloch equation (Macomber, 1976)). Since the integral of the system is $q_1^2 + q_2^2 + q_3^2 = r^2$, the motion can be regarded as occurring on the sphere S_2 of radius r in $\mathbf{R}^3$ with centre at the origin.

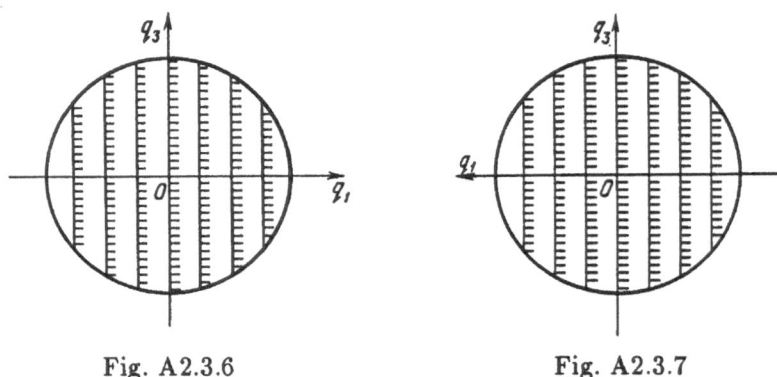

Fig. A2.3.6   Fig. A2.3.7

The family of trajectory funnel boundaries is the solution of the set of equations

$$\dot{q}_1 = 0, \ \frac{dq_3}{dq_2} = -\frac{q_3}{q_2} \tag{9}$$

and they are circles on the sphere lying in the planes

$$q_1 = C, \ |C| \le r. \tag{10}$$

The axis q_2 in Figure A2.3.6 is directed towards us. The axis q_2 in Figure A2.3.7 is directed away from us. The figures show the hemispheres that are closer to us and they are covered with the hatched families of trajectory funnel boundaries. It is clear from the figures that the system is completely controllable. For instance, starting from any point q_0 of the sphere it is possible to reach any point q_* making only one 'change', that is, first travelling along a trajectory (9) and then changing to a trajectory of free motion ($u = 0$) or vice versa.

In our discussion of the above examples we did not examine the explicit definition of the finite control

$$u = \bar{\phi}(t), \ 0 \le t \le T, \tag{11}$$

corresponding to the admissible trajectory $\xi(t)$, $0 \le t \le T$, $\xi(0) = q_0, \xi(T) = q_*$. However, the solution of this problem is not too difficult. Indeed, suppose that s is an admissible curve and $\bar{\xi}(a) = \xi$ is its parametrical equation with the parameter $a : a_0 \leq a \leq a_1$, while $\bar{\xi}(a_0) = q_0$ and $\bar{\xi}(a_1) = q_*$. Then according to the conditions for determining s, there exist for almost all $a \in [a_0, a_1]$ numbers $\lambda(a) > 0$ and $\bar{\phi}(a)$ such that

$$\lambda(a)\frac{\overline{d\xi}(a)}{da} = f(\bar{\xi}(a), \bar{\phi}(a)). \tag{12}$$

If for almost all $a \in [a_0, a_1]$ the integral

$$t = \int_{a_0}^{a} \frac{da}{\lambda(a)} \equiv \theta(a) \tag{13}$$

exists, then by replacing the independent variable a by the integral t in (12) we arrive at the equality

$$\frac{d\xi(\bar{\xi}(t))}{dt} = f(\bar{\xi}(\theta(t)), \ \bar{\phi}(\bar{\theta}(t))), \tag{14}$$

where $\bar{\theta}(t)$ is the inverse function of $\theta(a)$ defined by the equality (13). The function $\bar{\theta}(a)$ exists because $\theta(a)$ is monotonic in view of $\lambda(a) > 0$. It is also evident that the following boundary conditions hold:

$$\bar{\xi}(\bar{\theta}(0)) = q_0 \text{ and } \xi(\theta(T)) = q_{*1}, \quad T = \theta(a_1). \tag{15}$$

Thus, $u = \bar{\phi}(\bar{\theta}(t)) = \phi(t)$ and $q = \bar{\xi}(\bar{\theta}(t)) = \xi(t)$ is the required solution of the finite control problem.

We now consider the case where the dimension n in (1) of the vector $q = (q_1, \ldots, q_n) \in \mathbf{R}^n$ is greater than two. The problem is to find the equation of the hypersurface of the trajectory funnel with its vertex at the point q' of the differential inclusion (3). Denote this surface by $\mathcal{P}(q')$.

Suppose that $\mathcal{C}(q)$ is a cone with vertex at the point q; the cone is produced by all the rays emanating from the point q whose intersection with $f(q, \mathcal{U})$ is non-empty (the 'cone of admissible directions'). Also suppose that $\mathcal{C}(q)$ does not coincide with the whole of the space $\mathbf{R}^n$ at all points $q \in \mathbf{R}^n$ under consideration. Each cone $\mathcal{C}(q)$ is associated with the cone $\mathcal{C}'(q)$ with the vertex at the same point q and consisting of all the rays (emanating from the point q) such that the scalar product

$$(\tau, \tau') \leq 0, \tag{16}$$

where τ and τ' are arbitrary vectors originating in q, the ends of the vectors belonging to the cones $\mathcal{C}(q)$ and $\mathcal{C}'(q)$ respectively.

A relation essential for our subsequent derivation follows from this definition. Let p be a vector emanating from q and belonging to the surface of the cone $\mathcal{C}'(q)$. In other words, p is collinear with the generatrix of the cone $\mathcal{C}'(q)$. It is evident then that

$$\sup_{u \in \mathcal{U}} pf(q, u) = 0. \tag{17}$$

Supposing that the upper bound is attained, we can write

$$\max_{u \in \mathcal{U}} pf(q, u) = 0. \tag{18}$$

Let us call this relation the *maximum principle*.

We set

$$u(p, q) = \arg \max_{u \in \mathcal{U}} pf(q, u).$$

Then from (18) we have $pf(q, u(p, q)) = 0$, where the vector $f(q, u(p, q))$ is collinear with the generatrix of the cone $\mathcal{C}(q)$. We introduce the function $B(p, q, u) = pf(q, u)$. Now the relation above can be rewritten as

$$B(p, q, u(p, q)) = 0. \tag{19}$$

The idea of finding the equations of the surface $\mathcal{P}(q')$ consists in identifying the cone $\mathcal{C}(q)$ with the Monge cone (Tricome, 1958) of the generally non-linear scalar partial differential equation of the first order

$$F(p, q) = 0, \tag{20}$$

where

$$q = (q_1, \ldots, q_n), \quad p = \frac{\partial z}{\partial q} = \left(\frac{\partial z}{\partial q_1}, \ldots, \frac{\partial z}{\partial q_n} \right)$$

and $z = z(q)$ is an unknown scalar function. The sought surface $\mathcal{P}(q')$ is now given by the equation

$$z(q) = 0. \tag{21}$$

Usually, the equation (20) is specified, and then the Monge cone is found. Our problem is the converse: given the Monge cone, whose role is played by the cone $\mathcal{C}(q)$, recover equation (20), that is, define the form of the function $F(p, q)$.

A characteristic property of the required continuous piecewise-smooth surface $\mathcal{P}(q')$ is that at every 'smooth' point $q \in \mathcal{P}(q')$ the cone $\mathcal{C}(q)$ touches the surface $\mathcal{P}(q')$. The side of the surface $\mathcal{P}(q')$ where the cone $\mathcal{C}(q)$ touches it, will henceforth be called the *inner side*. Suppose that $p = \partial z/\partial q$ is normal to the surface $\mathcal{P}(q')$ at the point $q \in \mathcal{P}(q')$ which is on the surface of $\mathcal{C}'(q)$. It follows immediately that the equation needed in order to find the function $z(q)$ has the form

$$F\left(\frac{\partial z}{\partial q}, q\right) = B\left(\frac{\partial z}{\partial q}, q, u\left(\frac{\partial z}{\partial q}, q\right)\right) = 0. \tag{22}$$

Fixing the point $q' \in \mathbf{R}^n$ and, therefore, the cone $\mathcal{C}(q')$ defines the unique solution $z(q)$ of equation (22) which, in its turn, defines the surface we are seeking of the trajectory funnel of equation (3) with vertex at the point q'.

The partial differential equation (22) is associated with the following (canonical) system of ordinary differential equations of order $2n$, called the *characteristic system* for (22) because it defines the properties of the surface $z(q) = 0$ (Tricome, 1958):

$$\dot{q} = f(q, u(p, q)) = \frac{\partial B}{\partial p},$$
$$\dot{p} = -p\frac{\partial f}{\partial q}(q, u(p, q)) = -\frac{\partial B}{\partial q} \tag{23}$$

with the initial condition

$$q(0) = q', \ p(0) = p', \tag{24}$$

where the unit vector $p' \in \mathcal{C}(q')$.

The closed system of equations (23) and (24) gives the parametric equations of the required surface $\mathcal{P}(q')$ with two parameters: $t \geq 0$ and $p_o \in \mathcal{C}(q')$ ($|p_0| = 1$). Here $p = p(t, p_0)$ is the vector of the outward normal to $\mathcal{P}(q')$ at the point $q(t, p_0) \in \mathbf{R}^n$.

In consequence, the canonical (Hamiltonian) system (23) and the Jacobi equation (22) can be obtained as a corollary of the maximum principle (18). Unlike the maximum principle in optimal control and variational problems (N. Krasovskii, 1968), this one can be called the *maximum principle for the trajectory funnel* for the differential inclusion (3). Unlike the variational maximum principle which leads to the solution of the boundary problem for (23), this is an ordinary Cauchy problem whose solution presents no difficulties, at least in principle, with the use of computers.

Thus, we have used differential geometry to solve the general problem of finite control of the system (3). Let us formulate the sequence of operations in the process:

(1) Plot in $\mathbf{R}^n$ the family $\mathcal{S}$ of the surfaces (boundaries) of the trajectory funnels $\mathcal{P}(q')$, varying the parameter $q' \in \mathbf{R}^n$ so that these surfaces fill (cover) $\mathbf{R}^n$ or some specified part of it sufficiently densely.

(2) Hatch the outside of each surface $\mathcal{P}(q')$ of the family.

(3) Plot in $\mathbf{R}^n$ the line s from the given point q_0 to q_* so that this line is directed inside each funnel curve that it intersects.

(4) Identify the line s with a trajectory of the system (1).

These four operations guarantee that the sought finite control and the corresponding trajectory connects the given points q_0 and q_*.

Note that the examination of the family $\mathcal{S}$ may show that the operation cannot be carried out, and therefore the system is uncontrollable. Thus, there may exist a surface in the family $\mathcal{S}$ such that there can be no admissible trajectory passing from the surface unhatched side to the hatched side. This is the situation in the example 2 for the system (6). For instance, the hyperbola 1 divides $\mathbf{R}^2$ into two parts and is only hatched on one side (Figure A2.3.4). Therefore, there is no trajectory leading from the domain I to the domain II.

In conclusion, we note that the procedure described above, when n is more than 2, outlines the necessary conditions suggesting the existence of funnels and the absence of singular controls.

The idea of this section has been developed in greater detail in Butkovskiy (1982).

APPENDIX 3

Continuous Media and Controlled Dynamical Systems (CDS's). The Maximum Principle for Substance Flow. The Laplacian of a CDS

We show that each CDS defined by the inclusion $\dot{q} \in f(q, U)$, $q \in \{q\}$ can be associated with a certain continuous medium in which a certain 'substance', for example, heat, mass, charge, etc., is propagated. For definiteness, we shall consider thermal conduction occurring in a stationary medium in an n-dimensional space $\{q\} = \mathbf{R}^n$. The state of the medium will be given by the temperature distribution function $z = z(q, t)$, $q \in \{q\}$, where t is time. For brevity, we shall drop t in the function notation.

As is well known, the properties of such a medium are given by the operator $A(\mathrm{grad} z(q), q)$ that correlates the temperature gradient $\partial z / \partial q = \mathrm{grad}\, z(q)$ and the vector flow which we shall denote by $\kappa(q)$:

$$\kappa(q) = A(\mathrm{grad} z(q), q), \quad q \in \{q\}. \tag{1}$$

Relations of this type are called *material relations* in field theory. In the classical theory, the operator A is usually a linear algebraic operator relating $\kappa(q)$ and $\mathrm{grad}\, z(q)$.

However, there exist continuous media for which the material relations are not linear. These continuous media are called *non-linear*. Non-linear media occur in both nature and technology. Examples are laminated media, active media with negative conductivity and stored energy, and many other kinds of media (see Chapters 5 and 6).

Our aim will be, in particular, to make the relation (1) concrete, that is, to show that the quantities κ and $\mathrm{grad}\, z$ are related by the *maximum flow*

213

principle. Before we can formulate this principle, we have to introduce certain notions.

First we introduce the set $\Phi(q)$ of points of the tangent space $T_q\{q\}$. The set $\Phi(q)$ is given for each fixed $q \in \{q\}$. For the sake of simplicity we suppose that $\Phi(q)$ is a non-empty, bounded, closed, strictly convex set. Let us call it the *set of admissible flow directions.*

Strict convexity of $\Phi(q)$ is required here so that we can avoid any possible ambiguity in finding the flow $\kappa(q)$ from the maximum flow principle formulated below. It can be shown that for our purposes this requirement is not all that essential and can be replaced by simple convexity of $\Phi(q)$. However, we shall not deal with it here. The flow vector $\kappa(q)$ is assumed to occur in the space $T_q(q)$ and emanate from the point $\dot{O}(q)$. If the end of the vector $\kappa(q)$ belongs to $\Phi(q)$, then we write $\kappa(q) \in \Phi(q)$ or $\kappa \in \Phi(q)$.

The maximum flow principle. *Suppose that the vector $\kappa_m(q)$ satisfies the following maximum condition:*

$$\max_{\kappa \in \Phi(q)} [\operatorname{grad} z(q) \cdot \kappa] = \operatorname{grad} z(q) \cdot \kappa_m(q) \quad \forall q \in \{q\}, \tag{1}$$

or, what is the same,

$$\kappa_m(q) = \arg \max_{\kappa \in \Phi(q)} [\operatorname{grad} z(q) \cdot \kappa] \quad \forall q \in \{q\}. \tag{2}$$

Then the flow $\kappa(q)$ at the point $q \in \{q\}$, produced by $\operatorname{grad} z(q)$ equals

$$\kappa(q) = \alpha(|\operatorname{grad} z(q)|, q)\kappa_m(q), \tag{3}$$

where $\alpha(x,q)$ is a scalar function of the scalar non-negative argument $x \geq 0$, which (along with the set $\Phi(q)$) characterizes the thermal conductivity of the given non-linear continuous medium at each $q \in \{q\}$. The function α will be called the modulus function of the continuous medium. This is the formulation of the principle.

The suggested method of describing the material relation in non-linear continuous media can be justified, in particular, by the fact that the well-known material relations in linear media can easily be deduced from it. For instance, assume that $\Phi(q)$ is a set bounded by a hypersurface (indicatrix) in the form of an ellipsoid:

$$\frac{(q^1)^2}{a_1^2} + \ldots + \frac{(q^n)^2}{a_n^2} = 1, \tag{4}$$

where the lengths of the semi-axes $(a_i, \ i = 1, \ldots, n)$ may possibly depend on q. Now assume that $\alpha(x, q) \equiv |x|$. Then using (1), (2), and (3), it is easy to see by direct calculation that

$$\kappa(q) = [\operatorname{grad} z(q) \cdot B(q)]^T, \tag{5}$$

where $B(q)$ is a diagonal $n \times n$ matrix with generally distinct diagonal entries depending on $a_1, \ldots, a_n$.

When at least two of the $a_1, \ldots, a_n$ are unequal, it is well known that the relation (5) defines a linear continuous medium that is anisotropic with respect to its thermal conductivity. If in addition at least one of the numbers in question depends on $q \in \{q\}$, then the medium is heterogeneous. If, on the other hand, all the numbers (or functions of q) $a_1, \ldots, a_n$ are equal, then the medium is isotropic.

Let us discuss the relation (3) in more detail. It is clear from (3) that the flow $\kappa(q)$ depends non-linearly on $\operatorname{grad} z(q)$. The flow is the product of two factors: the vector $\kappa_m(q)$ and the scalar $\alpha(|\operatorname{grad} z(q)|, q)$. The vector factor specifies, so to speak, the non-linear dependence of $\kappa(q)$ on $\operatorname{grad} z(q)$ with respect to direction: it is seen in that the vector $\kappa(q)$ is generally non-collinear with the vector $\operatorname{grad} z(q)$. The only case when they are collinear for any $\operatorname{grad} z(q)$ is when $\Phi(q)$ is a sphere, that is, when the continuous medium is isotropic.

We further note that the zeroth power of $\kappa_m(q)$ is positively homogeneous with respect to the variable $\operatorname{grad} z(q)$, as is evident from (2). This implies that any multiplication of $\operatorname{grad} z(q)$ by a number $\lambda > 0$ in (2) does not change the value of $\kappa_m(q)$. In consequence, (1) and (2) can be respectively rewritten in the form (when $\alpha \neq 0$):

$$\max_{\kappa \in \Phi(q)} \left[\frac{\operatorname{grad} z(q)}{\alpha(|\operatorname{grad} z(q)|, q)} \kappa \right] = \frac{\operatorname{grad} z(q)}{\alpha(|\operatorname{grad} z(q)|, q)} \kappa_m(q), \tag{6}$$

$$\kappa_m(q) = \arg \max_{\kappa \in \Phi(q)} \left[\frac{\operatorname{grad} z(q)}{\alpha(|\operatorname{grad} z(q)|, q)} \kappa \right]. \tag{7}$$

The formulae (6) and (7) allow us to regard the vector $\kappa_m(q)$ as the *flow produced by a unit (normalized) temperature gradient.*

The scalar $\alpha(|\operatorname{grad} z(q)|, q)$ specifies the non-linear dependence of $\kappa(q)$ on $\operatorname{grad} z(q)$ with respect to the magnitudes of these vectors. This means that a λ-fold increase of $|\operatorname{grad} z(q)|$ $(\lambda > 0)$ does not generally increase $|\kappa(q)|$ in the same ratio. An exception is the case when $\alpha(x, q)$ for $x > 0$ is a proportional relation.

Therefore, equation (3) can be said to be a 'separated' non-linear dependence with two factors, one of which is responsible for the non-linearity with respect to the direction and the other with respect to the magnitude.

The introduction of α into (3) may seem to be questionable. However, this does not happen to be so. As experience and theory both show (Ladikov-Roev, 1978; Samoilenko, 1968), as well as Chapters 5 and 6 of this book, there are both natural and artificial media where a λ-fold increase in the magnitude of the vector ($\lambda > 0$) but with its direction kept constant does not entail an increase in $|\kappa(q)|$ in the same ratio. For instance, there are media where the electric current is not proportional to the electric field strength, that is, where Ohm's law, a *linear material relation*, is violated. The presence of such non-linear effects justifies the introduction of the modulus function α into (3).

We introduce the function $H(p, q)$ as the support function of the set $\Phi(q)$:

$$H(p, q) = \max_{\kappa \in \Phi(q)} p\kappa. \tag{8}$$

Then (1), (2), and (3) can be rewritten respectively as

$$H(p, q) = p\kappa_m(q), \tag{9}$$

$$\kappa_m(q) = \arg \max_{\kappa \in \Phi(q)} p\kappa = \frac{\partial H}{\partial p}(p, q), \tag{10}$$

$$\kappa(q) = \alpha(|p|, q)\frac{\partial H}{\partial p}(p, q), \tag{11}$$

where $p = \partial z(q)/\partial q = \operatorname{grad} z(q)$. Equation (11) defines a *non-linear* material relation in terms of the function $H(p, q)$.

It is easy to see the physical meaning of the function $H(p, q)$. In view of the positive homogeneity of the first power of $H(p, q)$, we have from (10):

$$p\kappa_m(q) = p\frac{\partial H}{\partial p}(p, q) = H(p, q). \tag{12}$$

But $p\kappa_m(q)$ indicates the quantity of heat flowing past q per unit time per unit area in the direction $p = \operatorname{grad} z(q)$ under the influence of a unit (normalized) gradient

$$\frac{p}{|p|} = \frac{\operatorname{grad} z(q)}{|\operatorname{grad} z(q)|}.$$

As (12) indicates, this quantity equals $H(p, q)$ when $p = \mathrm{grad}z(q)$. This is precisely the meaning of the function $H(p, q)$.

Similarly, we have from (11)

$$p\kappa(q) = \alpha(|p|, q)p\frac{\partial H}{\partial p}(p, q) = \alpha(|p|, q)H(p, q). \tag{13}$$

Equation (13) shows the rate of flow past q per unit area in the direction $p = \mathrm{grad}z(q)$ under the influence of $\mathrm{grad}\, z(q)$.

It is easy now to write down an expression for the Laplace operator $\Delta z(q)$ in the given non-linear continuous medium described in terms of $H(p, q)$ and the modulus function $\alpha(|p|, q)$. Indeed, by definition we have

$$\Delta z(q) = \mathrm{div}\, \kappa(q). \tag{14}$$

Substituting (11) into (14), we obtain the required relations for the Laplace operator in the given non-linear continuous medium:

$$\begin{aligned} \Delta z(q) &= \mathrm{div}\left[\alpha(|p|, q)\frac{\partial H}{\partial p}(p, q)\right], \\ p &= \frac{\partial z(q)}{\partial p} = \mathrm{grad}\, z(q). \end{aligned} \tag{15}$$

In coordinate form, (15) has the form

$$\Delta z(q) = \sum_{k=1}^{n} \frac{\partial}{\partial q^k}\left[\alpha(|p|, q)\frac{\partial H}{\partial p}(p, q)\Big|_{p=\partial z/\partial q}\right], \tag{16}$$

while in symbolic operator form it is

$$\begin{aligned} \Delta = \sum_{k=1}^{n} \frac{\partial}{\partial q^k}\left[\alpha(|p|, q)\frac{\partial H}{\partial p}(p, q)\Big|_{p=\partial(\cdot)/\partial q}\right] &= \\ = \frac{\partial}{\partial q}\left[\alpha(|p|, q)\frac{\partial H}{\partial p}(p, q)\Big|_{p=\partial(\cdot)/\partial q}\right]. \end{aligned} \tag{17}$$

The equation of thermal conductivity with respect to an unknown temperature distribution z can be written in terms of the Laplace operator. Thus the equation of the stationary temperature distribution has the form

$$\Delta z(q) = -\delta(q - q_0), \tag{18}$$

where the delta-function $\delta(q - q_0)$ defines the point heat source of unit power at $q_0 \in \{q\}$. Equation (18) can be regarded as the limiting case of the equation for non-stationary thermal conductivity (as $t \to \infty$):

$$\gamma \frac{\partial z}{\partial t} = \Delta z + g(q), \tag{19}$$

where γ is a coefficient and the function $g(q)$ defines the distribution of external stationary (that is, independent of time t) heat sources in $\{q\}$. In particular, in the case of a point source we have $g(q) = \delta(q - q_0)$. Here the temperature distribution $z = z(q, t)$ depends explicitly on t, and the distribution only becomes stationary, that is independent of time, in the limit as $t \to \infty$.

Now it is also not difficult to establish the relation between a CDS and a non-linear continuous medium of the above sort. Thus suppose that we are given a CDS defined by the equation

$$\dot{q} = f(q, u), \quad u \in U(q), \ q \in \{q\}, \tag{20}$$

or the inclusion

$$\dot{q} \in \Phi(q) \equiv f(q, U(q)), \tag{21}$$

and let the Hamiltonian of the CDS be

$$H(p, q) = \max_{u \in U(q)} pf(q, u). \tag{21a}$$

We identify the set $\Phi(q)$ of the given CDS with the set of admissible flow directions in the continuous medium. Therefore, the velocity $\dot{q}$ (see (21)) of the CDS is identified with the vector κ_m (see (2)). Similarly, the Hamiltonian $H(p, q)$ of the CDS is identified with the support function of the set of admissible flow directions. However, we are still free to ascribe a modulus function α to the given CDS (20), (21). When we have selected the modulus function α for the CDS, we therefore associate the CDS (20), (21) with the Laplace operator of the continuous medium (see (16), (17)). This then is how the continuous medium and its Laplace operator are related in terms of the given CDS.

Note that the freedom of choice of the modulus function α in establishing this correspondence may prove to be very useful for the solution of the reverse problem: given a continuous medium defined by a partial differential operator, associate it with a CDS defined by the relations (20), (21). Let us consider the reverse problem in more detail. Suppose that there is a differential expression (operator) $\psi(\partial^2 z/\partial q^2, \partial z/\partial q, q)$, where ψ is a given function of the variables $\partial^2 z/\partial q^2, \partial z/\partial q, q$. This operator can enter into the description of processes in a continuous medium. It is required to find a CDS defined by the relations (20), (21), (21a) (for example, find $H(p, q)$) that associates the continuous medium with the Laplace operator in the form of the operator ψ.

Obviously, the solution of the reverse problem can be reduced to the solution of the functional equation

$$\phi\left(\frac{\partial^2 z}{\partial q^2}, \frac{\partial z}{\partial q}, q\right) = \Delta z(q), \tag{22}$$

where $\Delta z(q)$ is defined by (16) (or (17)). As to (22), there are two unknown functions: $H(p, q)$ and $\alpha(|p|, q)$. The function $H(p, q)$ must be a positively homogeneous function of the first order in p, and the function α must be positive. If such solutions of (22) exist and can be determined, then the continuous medium described by the operator ψ is associated with a CDS of the form

$$\dot{q} = \frac{\partial H}{\partial p}(p, q), \tag{23}$$

where the vector p plays the role of control. It is clear that the freedom of choice of α in the solution of (22) helps one to find a function $H(p, q)$ satisfying the requirement of homogeneity laid down for it.

We consider a simple example of solving the reverse problem. Let

$$\psi\left(\frac{\partial^2 z}{\partial q^2}, \frac{\partial z}{\partial q}, q\right) \equiv \sum_{k=1}^{n} \frac{\partial^2 z}{(\partial q^k)^2}. \tag{24}$$

It is clear that in this case ψ can be represented in the form

$$\psi \equiv \Delta z \equiv \frac{\partial}{\partial z}\left(\frac{\partial z}{\partial q}\right) \equiv \operatorname{div}\frac{\partial z}{\partial q} \equiv \operatorname{div}\left[|\frac{\partial z}{\partial q}| \cdot \frac{\frac{\partial z}{\partial q}}{|\frac{\partial z}{\partial q}|}\right] \equiv \operatorname{div}\left[|p| \cdot \frac{p}{|p|}\right], \quad p = \frac{\partial z}{\partial q}. \tag{25}$$

In view of (25), it is easy to identify the functions $\alpha(|p|, q)$ and $H(p, q)$ possessing the required properties. The resulting solution is

$$\alpha(|p|, q) \equiv |p|, \quad \frac{\partial H}{\partial p} \equiv \frac{p}{|p|}, \quad H(p, q) \equiv |p|. \tag{26}$$

Therefore, the required equation of the CDS corresponding to the continuous medium with the operator (24) has the form

$$\dot{q} = \frac{\partial H}{\partial p} = \frac{p}{|p|}, \tag{27}$$

where p plays the role of control.

Since $H(p, q)$ is the support function of the convex set $f(q, U(q)) \equiv \Phi(q)$, it is easy to recover the set itself: it is the circle of radius 1 with centre at the point $\dot{O}(q)$. Therefore the CDS equation in terms of the control u has the form

$$\dot{q} = u, \quad u \in U = \{u : |u| \leq 1\}. \tag{28}$$

The CDS (28) is very simple: the entire state space $\{q\} = \mathbf{R}^n$ is the domain of its free motion since

$$\min_{|p|=1} H(p, q) = \min_{|p|=1} |p| = 1 > 0 \quad \forall q \in \{q\}, \tag{29}$$

where $H(p, q)$ is defined by (8).

In conclusion we discuss the significance of the results obtained here. It is well to bear in mind that the procedure of establishing a correspondence between a CDS and a process in a continuous medium can prove to be useful when studying a process or a system by means of another. For instance, problems formulated for a continuous medium can be reformulated as appropriate problems for a CDS, and *vice versa*. Thus, the problem of CDS controllability can be solved in terms of thermal conductivity. In fact, suppose that the given CDS is associated with the continuous medium possessing the Laplace operator $\Delta z(q)$ in the form (16) or (17). Suppose it is established that equation (18) of the stationary process of thermal conductivity has the solution $z(q)$ that is non-zero for every $q \in \{q\}$. Then it can be concluded that the given CDS is controllable in $\{q\}$. However, if $z(q) = 0$ at some points, the conclusion is that the given CDS is uncontrollable.

References

Aizerman, M.A.: 1980, *Classical Mechanics*, Nauka, Moscow (Russian).

Akhiezer, A.I., V.G. Bariyakhtar, and S.V. Peletminskii: 1970, *Spin Waves*, Nauka, Moscow (Russian).

Akhiezer, A.I., and V.B. Berestetskii: 1969, *Quantum Electrodynamics*, Nauka, Moscow (Russian).

Alekseev, O.V.: 1968, *Power Amplifiers with Distributed Amplification*, Energiya, Leningrad (Russian).

Alimpiev, S.S., N.V. Karlov, B.B. Krynetskii, and Yu.N. Petrov: 1980, 'Laser Separation of Isotopes', in *Achievements in Science and Technology. Radiotechnical Series*, Part I, VINITI, Moscow (Russian).

Allen, L., and J. Eberly: 1975, *Optical Resonance and Two-Level Atoms*, Wiley, New York.

Andreev, N.I.: 1980, *Theory of Statistically Optimal Control Systems*, Nauka, Moscow (Russian).

Andreev, Yu.N.: 1976, *Control of Finite-Dimensional Linear Plants*, Nauka, Moscow (Russian).

Andreev, Yu.N.: 1977, 'Algebraic Methods in the Space of States and the Theory of Control of Linear Plants', *Avtom. i Telem.* **3**.

Andreev, Yu.N.: 1982, 'Differential Geometry Methods in the Theory of Control (a Review)', *Avtom. i Telem.* **9**.

Aranjo Cid, B., and S.M. Rezende: 1974, 'Saturation and Coherence Properties of Three-Magnon Non-Linear Processes', *Physical Review*, Ser. B. **9** April No. 7, 321–335.

Artemenkov, L.I., I.N. Golovin, P.I. Kozlov, P.I. Melikhov, N.N. Shvindt, V.K. Butenko, V.F. Gubarev, A.I. Kukhtenko, Yu.P. Ladikov-Roev, and Yu.I.

Samoilenko: 1971, 'Experiments on Equilibrium in the Uncased TO-I Toka-
mak Device by Means of a Feedback System', in *Plasma Physics and Controlled Nuclear Fusion Research* **1**, International Atomic Energy Agency,
Vienna, 359–367.

Baitman, M.N.: 1978, 'On Controllability Domains in a Plane', *Differentsialnye
Uravneniya* **4**.

Basov, N.G., and A.M. Prokhorov: 1954, 'The Use of Molecular Beams for the
Study of Molecular Rotational Spectra', *Zh. Teoret. Fiz.* **27**:4 (10).

Baz, A.I., Ya.B. Zeldovich, and A.M. Perelomov: 1971, *Scattering, Reactions,
and Decay in Non-Relativistic Quantum Mechanics*, Nauka, Moscow (Russian).

Begimov, I., A.G. Butkovskiy, and V.L. Rozhanskii: 1980, 'Simulation of Complicated Distributed Systems Underlain by their Structural Theory', I,II,
Avtom. i Telem. **11,12**.

Begimov, I., A.G. Butkovskiy, and V.L. Rozhanskii: 1982, 'Structural Representation of Heterogeneous Physical Systems', *Avtom. i Telem.* **8**.

Belavkin, V.P.: 1974, 'Optimal Linear Randomized Filtration of Quantum Bo-
son Signals', *Problems of Control and Information Theory* **3**:1, 47–62.

Belavkin, V.P.: 1975, 'Optimization in Quantum Signal Processing (Review)',
Zarubezhnaya Elektronika **5**.

Belavkin, V.P.: 1980, 'Quantum Filtration of Markov Signals against the Back-
ground of White Quantum Noise', *Radiotekh. i Elektron.* **25**:7, 1446–1453.

Bellmann, R.: 1957, *Dynamic Programming*, Princeton Univ. Press, Princeton,
New York.

Blokhintsev, D: 1981a, *Principes de mécanique quantique*, Mir, Moscow.

Blokhintsev, D.I.: 1981b, *Quantum Mechanics*, Atomizdat, Moscow (Russian).

Bogolyubov, N.N., and Yu.A. Mitropolskii: 1963, *Asymptotic Methods in the
Theory of Non-Linear Oscillations*, Fizmatgiz, Moscow (Russian).

Bogolyubov, N.N., and D.V. Shirokov: 1976, *Introduction to the Theory of
Quantum Fields*, Nauka, Moscow (Russian).

Boltyanskii, V.G.: 1966, *Mathematical Methods of Optimal Control*, Nauka,
Moscow (Russian).

Bradshaw, A.: 1974, 'Modal Control of Distributed Parameter Vibratory Systems', *Int. J. Control* **19**:5, 957–968.

Bradshaw, A., and B. Porter: 1972, 'Modal Control of a Class of Distributed

Parameter Systems: Multi-eigenvalue Assignment', *Int. J. Control.* **16**:2, 277–285.

Braginskii, V.B.: 1970, *Physical Experiments with Test Bodies*, Nauka, Moscow (Russian).

Bredly, D.J.: 1977, 'The Laser: the Dynamics of the Twenty-First Century', *J. Royal Society Arts*, Nov., 3–20.

Brockett, R.W.: 1972, 'System Theory on Group Manifolds and Coset Spaces', *SIAM J. Control* **10**:2, 265–284.

Brockett, R.W.: 1973a, 'Lie Algebras and Lie Groups in Control Theory', in *Geometrical Methods in Systems Theory*, Reidel, Dordrecht, 43–88.

Brockett, R.W.: 1973b, 'Lie Theory and Control Systems Defined on Spheres', *SIAM J. Applied Math.* **25**:2, 213.

Brockett, R.W., and A.S. Willsky: 1975, 'Some Structural Properties of Automata Defined on Groups', *Lecture Notes in Computer Science*, 1-75, No.25, 112–118.

Bulatov, V.P.: 1977, *Immersion Methods in Optimization Problems*, Nauka, Novosibirsk (Russian).

Bunkin, F.V., N.A. Kirichenko, and B.S. Lukiynxhuk: 1978, 'Optimal Modes of Laser Heating of Materials', Preprint *Fiz. Inst. Akad. Nauk* No.146 (Russian).

Butkovskiy, A.G.: 1965, *Theory of Optimal Control for Distributed-Parameter Systems*, Nauka, Moscow (Russian).

Butkovskiy, A.G.: 1969, *Distributed Control Systems*, Elsevier, New York.

Butkovskiy, A.G.: 1975, *Methods of Control of Distributed-Parameter Systems*, Nauka, Moscow (Russian).

Butkovskiy, A.G.: 1977, *Structural Theory of Distributed Systems*, Nauka, Moscow (Russian).

Butkovskiy, A.G.: 1979a, 'Control of Distributed-Parameter Systems (Review)', *Avtom. i Telem.* **11**.

Butkovskiy, A.G.: 1979b, *Characteristics of Distributed-Parameter Systems*, Nauka, Moscow (Russian).

Butkovskiy, A.G.: 1982, 'Differential Geometry Technique of Structural Solutions for Problems of Controllability and Finite Control', *Avtom. i Telem.* **1**.

Butkovskiy, A.G., and L.M. Pustylnikov: 1980, *Theory of Flexible Control of*

Distributed-Parameter systems, Nauka, Moscow (Russian).

Butkovskiy, A.G. and E.I. Pustylnikova: 1982, 'Control of Coherent States of Quantum Oscillators', *Avtom. i Telem.* **11**.

Butkovskiy, A.G., and Yu.I. Samoilenko: 1979a, 'Control of Quantum Processes and Systems', Part 1, *Avtom. i Telem.* **5**.

Butkovskiy, A.G., and Yu. I. Samoilenko: 1979b, 'Control of Quantum Processes and Systems', Part II, *Avtom. i Telem.* **5**.

Butkovskiy, A.G., and Yu.I. Samoilenko: 1980, 'Controllability of Quantum Processes and Systems', *Dokl. Akad. Nauk SSSR* **250**:1.

Chernousko, F.L., and V.B. Kolmanovskii: 1978, *Optimal Control in Random Perturbations*, Nauka, Moscow (Russian).

Coddington, E.A., and N. Levinson: 1953, *Theory of Ordinary Differential Equations*, McGraw-Hill, New York.

Damon, R.W., and T.I. Eshbach: 1960, *Transactions of the IREE*, MIT **80**:4.

Davydov, A.S.: 1973, *Quantum Mechanics*, Nauka, Moscow (Russian).

Deryugin, I.A., P.S. Kuts, and V.L. Strizhevskii: 1967, 'Ferrite UHF-Modulators in Ferromagnetic Resonance', *Radiotekhnika* **22**:6.

Dirac, P.A.M.: 1958, *The Principles of Quantum Mechanics*, Clarendon Press, Oxford, 4th ed.

Dixmier, J.: 1974, *Algèbres enveloppantes*, Gauthier-Villars, Paris.

Dodonov, V.V., V.I. Manko, and V.N. Rudenko: 1980, 'Non-Perturbing Measurement in a Gravitation-Wave Experiment', *Zh. Eksper. Teoret. Fiz.* **8**:3.

Dubrovin, B.A., S.P. Novikov, and A.T. Fomenko: 1979, *Modern Geometry*, Nauka, Moscow (Russian).

Elsgolts, L.E.: 1957, *Differential Equations*, Gostekhizdat, Moscow (Russian).

Emeliyanov, S.V.: 1967, *Automatic Control Systems with a Variable Structure*, Nauka, Moscow (Russian).

Fabrikov, V.A.: 1958, 'A Possibility of Using Gyromagnetic Media to Amplify a Weak Modulating Signal', *Radiotekh. i Elektron.* **2**.

Fabrikov, V.A.: 1961, 'A New Principle of Using Ferrites to Amplify UHF Oscillations', *Radiotekh. i Elektron.* **2**.

Feldbaum, A.A.: 1971, *Fundamentals of the Theory of Optimal Automatic Systems*, Nauka, Moscow (Russian).

Feldbaum, A.A., and A.G. Butkovskiy: 1971, *Methods of the Theory of Automatic Control*, Nauka, Moscow (Russian).

Feynman, R.P., and A. Hibbs: 1965, *Quantum Mechanics and Path Integrals*, McGraw-Hill, New York.

Feynman, R.P., R.B. Leighton, and M. Sands: 1963, *The Feynman Lectures on Physics* Vol 9, Addison Wesley, Reading, Mass.

Fleming, W., and R. Richel: 1975, *Deterministic and Stochastic Optimal Control*, Springer, Berlin.

Flügge, S: *Practical Quantum Mechanics*, Vols 1–2, Springer, Berlin.

Gabasov, R., and F.M. Kirillova: 1971, *Qualitative Theory of Optimal Processes*, Nauka, Moscow (Russian).

Gantmakher, F.R.: 1966, *Theory of Matrices*, Nauka, Moscow (Russian).

Gertsenshtein, M.E.: 1959, 'Quasilinear Susceptibility of Ferrites to a Pumping Signal', *Radiotekh. i Elektron.* **11**.

Gertsenshtein, M.E., F.A. Levinson, A.A. Belov, and B.A. Tetelbaum: 1971, 'Three-Frequency Parameter System as a Negative Capacitor', *Radiotekh. i Elektron.* **16**:6.

Girko, V.L., and A.V. Vinogradskaya: 1979, 'Spectrum Control of Linear Operators in a Hilbert Space', *Vyshisl. Prikl. Mat.*, **38**.

Glizer, V., M.G. Dmitriev, and I.V. Krasnov: 1979, 'Optimization of Resonance Fluorescence', in *Proc. All-Union Conference on Dynamic Control*, Sverdlovsk (Russian).

Gurevich, A.G.: 1973, *Magnetic Resonance in Ferrites and Antiferromagnets*, Nauka, Moscow (Russian).

Hartman, P.: 1964, *Ordinary Differential Equations*, Wiley, New York.

Helstrom, C..W.: 1976, *Quantum Detection and Estimation Theory*, Academic Press, New York.

Irodov, I.E.: 1978, *Collection of Problems in Atomic and Nuclear Physics*, Atomizdat, Moscow (Russian).

Jurdjevič, V., and H.J. Sussman: 1972, 'Control Systems on Lie Groups', *J. Differential Equations* **12**, 313–329.

Karlov, N.V., and T.I. Kuznetsova: 1967, 'Non-Linear Distortion of Signals in Saturated Paramagnetic Amplifiers', *Radiotekh. i Elektron.* **12**:2.

Karlov, N.V., and A.A. Manenkov: 1966, 'Quantum Amplifiers', in *Itogi Nauki. Ser. Fiz. Inst. Nauchn. Inf. Akad. Nauk SSSR*, Moscow.

Kholevo, A.S.: 1980, *Probabilistic and Statistical Aspects of Quantum Theory*, Nauka, Moscow (Russian).

Khorozov, O.A.: 1980, 'The State of a Quantum System at the Maximum Observation Probability', in *Proc. 8th All-Union Conference of USSR SCAC*, Tallinn (Russian).

Kielich, S.: 1977, *Molekularna Optyka Nieliniowa*, Warszawa-Poznan.

Klauder, J., and E. Sudarshan: 1968, *Fundamentals of Quantum Optics*, Benjamin, New York.

Kostrikin, A.I.: 1978, *Introduction to Algebra*, Nauka, Moscow (Russian).

Krasnoselskii, M.A.: 1956, *Topological Methods in the Theory of Non-Linear Differential Equations*, Gostekhizdat, Moscow (Russian).

Krasnov, I.V., N.Ya. Shaparev, and N.M. Shkedov: 1978, 'Optimal Control of Photoprocesses in Gases', *Vyschisl. Sistemy*, Preprint No.10, Krasnoyarsk (Russian).

Krasovskii, A.A.: 1968, *The Statistical Theory of Transition Processes in Control Systems*, Nauka, Moscow (Russian).

Krasovskii, A.A.: 1974, 'Ultimate Accuracy in Micro-Observation and Micro-Control', *Izv. Akad Nauk SSSR, Tekhn. Kibernet.* **3**.

Krasovskii, A.A., and G.S. Pospelov: 1962, *Fundamentals of Automation and Technical Cybernetics*, Gosenergoizdat, Moscow (Russian).

Krasovskii, N.N.: *The Theory of Motion Control*, Nauka, Moscow (Russian).

Krein, S.G. (ed): 1964, *Reference Mathematical Library. Functional Analysis*, Nauka, Moscow (Russian).

Kubo, R.: 1957, 'Statistical-Mechanical Theory of Irreversible Processes. I. General Theory and Simple Applications to Magnetic and Conduction Problems', *J. Phys. Soc. Japan* **12**, 570–586.

Kuchtenko, A.I., and Yu. I. Samoilenko: 1973, 'Stabilisiertes Plasma für die Kontrolierte Kernfusion. Ideen des exakten Wissens', *DVA*, Stuttgart, Juli, **7**, 413–422.

Kukharkin, E.S.: 1969, *Fundamentals of Electrophysics for Engineers*, Part I, Vysshaya Shkola, Moscow (Russian).

Kuriksha, A.A.: 1973, *Quantum Optics and Laser Radars*, Sov. Radio, Moscow (Russian).

Kurzhanskii, A.B.: 1977, *Control and Observation Under Uncertainty Conditions*, Nauka, Moscow (Russian).

Kuts, P.S., and V.S. Melnik: 1974, 'Dynamic Phenomena in Magnetic Resonance whereas a Gyromagnetic Medium is Magnetized Harmonically', *Zh.*

Teoret. Fiz. **44**:1.

Kuts, P.S., V.S. Melnik, and V.L. Strizhevskii: 1973, 'Dynamics of Magnetization in Harmonically Modulated Gyromagnetic Media', *Radiotekh. i Elektron.* **18**:7.

Ladikov-Roev, Yu.P.: 1978, *Control of Continuous Media*, Nauka, Moscow (Russian).

Landau, L.D., and I.M. Lifshits: 1935, 'On the Theory of the Dispersion of Magnetic Permeability on Ferromagnetic Bodies', *Phys. Zeitschrift Soviet Junion* **8**, 153.

Landau, L.D., and E.M. Lifshits: 1959, *Electrodynamics of Continuous Media*, Fizmatgiz, Moscow (Russian).

Landau, L.D., and E.M. Lifshits: 1963, *Quantum Mechanics. Non-Relativistic Theory*, Fizmatgiz, Moscow (Russian).

Landau, L.D., and E.M. Lifshits: 1965, *Mechanics*, Nauka, Moscow (Russian).

Landau, L.D., and E.M. Lifshits: 1976, *Statistical Physics*, Part I, Nauka, Moscow (Russian).

Lappo-Danilevsky, I.A.: 1957, *Application of Matrix Functions to the Theory of Linear Systems of Ordinary Differential Equations*, Gostekhteorizdat, Moscow (Russian).

Lavrentiev, M.A., and B.V. Shabat: 1965, *Methods of Function Theory*, Nauka, Moscow (Russian).

Lax, B., and K. Button: 1962, *Microwave Ferrites and Ferromagnetics*, McGraw-Hill, New York.

Lions, J.L.: 1968, *Contrôle optimal de systèmes gouvernés par des équations aux dérivées partielles*, Dunod Gauthier-Villars, Paris.

Lurie, K.A.: 1975, *Optimal Control in Problems of Mathematical Physics*, Nauka, Moscow (Russian).

L'vov, V.S., and S.S. Starobinets: 1971, 'On Non-Linear Theory of Ferromagnetic Resonance', *Fizika Tverd. Tela* **13**:2.

Lyubarskii, G.Ya.: 1958, *Group Theory and its Application in Physics*, Fizmatgiz, Moscow (Russian).

Mackey, G: 1962, *The Mathematical Foundations of Quantum Mechanics. A Lecture-Note Volume*, New York.

Macomber, J.D.: 1976, *The Dynamics of Spectroscopic Transitions*, Wiley, New York.

Malkin, I.A., and V.I. Manko: 1979, *Dynamic Symmetries and Coherent States of Quantum Systems*, Nauka, Moscow (Russian).

Merkin, D.R.: 1971, *Introduction to the Theory of Motion Stability*, Nauka, Moscow (Russian).

Merkin, D.R.: 1974, *Gyroscopic Systems*, Nauka, Moscow (Russian).

Migdal, A.B.: 1975, *Qualitative Methods in Quantum Mechanics*, Nauka, Moscow (Russian).

Mikhaelyan, A.L.: 1963, *The Theory and Application of Ferrites at Ultra-High Frequencies*, Gosenergoizdat, Moscow (Russian).

Moiseev, N.N.: 1971, *Numerical Methods in the Theory of Optimal Systems*, Nauka, Moscow (Russian).

Moncrief, V.: 1978, 'Coherent States and Quantum Non-Perturbing Measurements', *Ann. Phys.* **114**, 201–214.

Monosov, A.Ya.: 1970, *Non-Linear Ferromagnetic Resonance*, Nauka, Moscow (Russian).

Morgenthaller, F.R.: 1960, 'Survey on Ferromagnetic Resonance in Small Ferromagnetic Ellipsoids', *J. Appl. Phys.* **31**, 555–959.

Mustafaev, M.I.: 1981, 'Fundamental Finite Control in Source-like Perturbation', *Avtom. i Telem.* **12**.

Neumann, J. von: 1932, *Mathematische Grundlagen der Quantenmechanik*, Springer, Berlin.

Nikityuk, N.M.: 1967, *Light, Quanta, and Computer Technology*, Znanie, Moscow (Russian).

Nikolai, E.L.: 1952, *Theoretical Mechanics*, Part 2, Gostekhizdat, Moscow (Russian).

Osborne, I.A.: 1945, 'Diamagnetizing Factors of the General Ellipsoid', *Phys. Rev.* **67**, June, 196, 351–357.

Ovsyannikov, L.V.: 1978, *Group Analysis of Differential Equations*, Nauka, Moscow (Russian).

Pantell, R.H., and H.E. Puthoff: 1969, *Fundamentals of Quantum Electronics*, Wiley, New York.

Patton, C.E.: 1969, 'Theory of the First Order Spin Waves Instability Threshold in Ferromagnetic Insulator of Ellipsoidal Shape with an Arbitrary Pumping Infiguration', *J. Appl. Phys.* **40**, 2837–2840.

Petrov, B.N.: 1961, 'Invariance Principle in the Calculation of Linear and Non-

Linear Systems', in *Proceedings of the 1st Congress of IFAC*, Izdat. Akad. Nauk SSSR, Moscow (Russian).

Petrov, B.N., G.M. Ulanov, and I.I. Goldenblat: 1978, *Theory of Simulation in Control Processes (the Aspects of Information and Thermodynamics)*, Nauka, Moscow (Russian).

Petrov, B.N., G.M. Ulanov, I.I. Goldenblat, I.D. Kochubievskii, E.M. Khazen, and S.V. Ul'yanov: 1976, 'Information Aspects of the Qualitative Theory of Dynamic Systems', in *Achievements in Science and Technology. Technical Cybernetics*, VINITI, Moscow, Vol. 7 (Russian).

Petrov, V.V., and A.S. Uskov: 1975, *The Information Theory of Synthesizing Optimal Systems for Measurement and Control*, Energiya, Moscow (Russian).

Pomerantsev, N.M., V.M. Ryzhkov, and G.V. Skrotskii: 1972, *Physical Foundations of Quantum Magnetometry*, Nauka, Moscow (Russian).

Pontryagin, L.S., V.G. Boltyanskii, R.V. Gamkrelidze, and E.F. Mishchenko: 1963, *Mathematical Theory of Optimal Processes*, Fizmatgiz, Moscow (Russian).

Popov, E.P.: 1954, *Dynamics of Systems of Automatic Regulation*, Gostekhizdat, Moscow (Russian).

Pugachev, V.S.: 1957, *The Theory of Probability Functions and Problems of Automatic Control*, Gostekhizdat, Moscow (Russian).

Pugachev, V.S., I.E. Kazakov, and L.G. Evlanov: 1974, *Fundamentals of the Statistical Theory of Automatic Systems*, Mashinostroenie, Moscow (Russian).

Ray, W..H., and D.G. Lainiotis (eds): 1978, *Distributed Parameter Systems*, Marcel Dekker, New York.

Razmakhin, M.K., and V.P. Yakovlev: 1971, *Functions with Double Orthogonality in Radioelectronics and Optics*, Sov. Radio, Moscow (Russian).

Roitenberg, Ya.N.: 1971, *Automatic Control*, Nauka, Moscow (Russian).

Rozenvasser, E.N.: 1973, *Periodically Non-Stationary Control Systems*, Nauka, Moscow (Russian).

Rozonoer, L.I.: 1959, 'The Pontryagin Maximum Principle in the Theory of Optimal Systems', *Avtom. i Telem.* **10,11,12**.

Rummer, Yu.V., and A.I. Fet: 1970, *The Theory of Unitary Symmetry*, Nauka, Moscow (Russian).

Samoilenko, Yu.I.: 1968, 'Space-Distributed Automatic Control Systems and Methods of their Realization', *Avtom. i Telem.* **2**.

Samoilenko, Yu.I.: 1972, 'Electromagnetic Control of Charged Particles with a Special Reference to Statistical and Quantum Effects. Proceedings of Kiev Summer School, 1971', in *Controllable Statistical Processes and Systems*, Izdat. Inst. Kibernet. Akad. Nauk USSR, Kiev, pp. 120–140 (Russian).

Samoilenko, Yu.I.: 1973, 'A Device for Confining of Plasma. Author's Certificate No. 356979', *Byulleten Izobret.* **9**.

Samoilenko, Yu.I.: 1974, 'Parametrically Induced Superdiamagnetism and its Future for Distributed Control', in *Distributed Control of Processes in Complicated Media*, Izdat. Inst. Kibernet. Akad. Nauk USSR, Kiev (Russian).

Samoilenko, Yu.I.: 1978a, 'The Phenomenon of Negative Susceptibility in Dynamic Systems and its Application to Stabilize Non-Stationary Processes and Systems', *Avtom. i Telem.* **5**.

Samoilenko, Yu.I.: 1978b, 'Realization of Rapid Negative Feedback Using Dynamic Systems with Stored Energy', *Avtom. i Telem.* **12**.

Samoilenko, Yu.I.: 1982a, 'Optimal Control of a Quantum Statistical Ensemble', *Avtom. i Telem.* **12**.

Samoilenko, Yu.I.: 1982b, 'Control on Groups of Quantum System Motion', in *Republican Interbranch Collection "Controlled Processes and Systems"*, Izdat. Kievsk. Gos. Univ., Kiev (Russian).

Samoilenko, Yu.I., and O.A. Khorozov: 1977, 'Stability of Spin Waves in Longitudinal Pumping of Ferromagnetic Crystals' in *Control of Distributed-Parameter Plants*, Izdat. Inst. Kibernet. Akad. Nauk USSR, pp. 80–88 (Russian).

Samoilenko, Yu.I., and O.A. Khorozov: 1980, 'A Possibility of Designing Digital Automata on Controlled Transitions of Quantum Systems', *Dokl. Akad. Nauk USSR* **11**.

Schlömann, E: 1964, 'The Growth of Spin Waves in Parametrical Excitation during the Transition Period', *Izv. Akad. Nauk SSSR, Ser. Fiz.* **3** (Russian).

Schlömann, E., G. Green, and V. Milano: 1960, 'Recent Development of Ferromagnetic Resonance of High Power Level', *J. Appl. Phys.* **31**, 3865–3871.

Sedov, L.I.: 1976, 'Trends in Problems of the Mechanics of Continuous Media', MPP 40:6 (Russian).

Sirazetdinov, T.K.: 1977, *Optimization of Distributed-Parameter Systems*,

Nauka, Moscow (Russian).

Skrotskii, G.V., and L.V. Kurbatov: 1961, 'Phenomenological theory of Ferromagnetic Resonance', in *Ferromagnetic Resonance*, Fizmatgiz, Moscow (Russian).

Schlichter, C.P.: 1980, *Principles of Magnetic Resonance*, Springer, Berlin.

Solodov, A.V., and F.S. Petrov: 1971, *Linear Systems with Variable Parameters*, Nauka, Moscow (Russian).

Sultanov, I.A.: 1980, 'Control of Processes Described by Equations with Indeterminate Functional Parameters', *Avtom. i Telem.* 10.

Tarasov, L.V.: 1976, *Physical Foundations of Quantum Electronics*, Sov. Radio, Moscow (Russian).

Thomson, W., and P. Tait: 1879, *Treatise on Natural Philosophy*, Part I, Cambridge Univ. Press, London.

Tikhonov, A.N., and V.Ya. Arsenin: 1979, *Methods of Solving Ill-Posed Problems*, Nauka, Moscow (Russian).

Tricome, F.G.: 1958, *Equazioni a derivate parzioli*, Ed. Cremonese, Rome.

Tsypkin, Ya.Z: 1977, *Foundations of the Theory of Automatic Systems*, Nauka, Moscow (Russian).

Unruh, W.: 1977, *An Analysis of Quantum-Non-Demolition Measurement* (Preprint), University of British Columbia.

Utkin, V.I.: 1974, *Shifting Modes and their Application to Systems with Variable Structure*, Nauka, Moscow (Russian).

Van der Heide, H.: 1976, 'Stabilization by Oscillation', *Philips Tech. Rev.* 34:2/3, 61–72.

Venikov, G.V.: 1976, *Optoelectronic Computer Systems*, Znanie, Moscow (Russian).

Vinogradskaya, A.V., and V.L. Girko: 1980, 'Control of the Spectrum in Systems Defined by Linear Equations in a Hilbert Space', *Avtom. i Telem.* 3.

Vonsovskii, S.V.: 1971, *Magnetism*, Nauka, Moscow (Russian).

Voronov, A.A.: 1966, *Fundamentals of the Theory of Automatic Control*, Energiya, Moscow (Russian).

Voronov, A.A.: 1979, *Stability, Controllability and Observability*, Nauka, Moscow (Russian).

Vorontsov, M.A.: 1978, 'Synthesis of Optimal Control of a Class of Distributed-Parameter Systems with Statistical Perturbations', *Izv. Akad. Nauk SSSR, Tekhn. Kibernet.* 5.

Vorontsov, Yu.I.: 1981, 'The Uncertainty Relation between Energy and Time of Measurement', *Uspekhi Fiz. Nauk* 133:2.

Waugh, J.S.: 1976, *New NMR Methods in Solid State Physics*, MIT, Cambridge, Mass.

Wigner, E.: 1959, *Group Theory and its Application to the Quantum Mechanics of Atomic Spectra*, Academic Press, New York.

Zaezdny, A.M., and V.F. Kushnir: 1962, *Parametrical systems*, Izdat. Leningrad. Electrotekh. Inst. Svyazi, Leningrad.

Zubarev, D.N.: 1971, *Non-Equilibrium Statistical Thermodynamics*, Nauka, Moscow (Russian).

Zubov, V.I.: 1975, *Lectures on the Theory of Control*, Nauka, Moscow (Russian).